Timber Wolves

Greed and Corruption in Northwestern Ontario's Timber Industry, 1875-1960

by

J.P. Bertrand

THUNDER BAY
MUSEUM

Publisher: The Thunder Bay Historical Museum Society Inc.
425 Donald St. E., Thunder Bay, Ont. P7E 5V1
tel: (807) 623-0801; fax: (807) 622-6880; e-mail: tbhms@tbaytel.net

Cover and page design: Thorold J. Tronrud

Printed by: Friesens Corporation
One Printer's Way
Altona, Manitoba R0G 0B0
Canada

Printed on acid-free paper

Canadian Cataloguing in Publication Data

Bertrand, J.P.
    Timber wolves : greed and corruption in northwestern Ontario's timber industry, 1875-1960

Includes index.
ISBN 0-920119-26-3

1. Lumber trade - Ontario, Northern - History. 2. Forest products industry - Ontario, Northern - History.
I. Thunder Bay Historical Museum Society. II. Title.

HD9764.C33057 1997   338.1'74980'971311   C97-931675-8

Dedicated

to

James Arthur Mathieu,
pioneer lumber manufacturer in Northwestern Ontario

Contents

Maps and Illustrations

Maps:

Illustrations:

Editor's Preface

Joseph Placide Theodore Bertrand (1880-1964) was born on a farm in the Ottawa valley near St. Placide Quebec. In 1900, he moved to Thunder Bay where he proceeded to build a career for himself in the lumber industry. He travelled frequently over the length and breadth of Northwestern Ontario developing a full appreciation for the unique role the region played in the history of the nation.

J.P., as he was popularly known, was an avid reader with a life-long passion for history. He also had a decided talent for writing, as his first book, *Highway of Destiny*, a best-selling history of Northwestern Ontario from the beginning of European settlement in the eighteenth century to the 1880s, attests. He was a meticulous researcher at a time when sources for the study of history in Northwestern Ontario were sparce and difficult to access. *Highway of Destiny* was the first serious attempt to record the early story of Northwestern Ontario.

This, his second book, carries the region's story forward into the twentieth century with a special focus on the forest industry, a business he knew most intimately. Having worked for the Pigeon River Lumber Company through much of its early development and having been, himself, a witness at the famous Timber Commission inquiry of 1920, Bertrand developed a close working relationship with many of the people described in this book, making *Timber Wolves* as much a primary document of Bertrand's working career as the work of an historian.

J.P. Bertrand served as president of the Port Arthur Rotary Club and headed the city's centennial committee in 1957. He was a past president of and significant contributor to the Thunder Bay Historical Society. The Society, in 1963, made him an honourary life member in recognition of his contributions to the Society and the history of Northwestern Ontario. He was also made an honourary chief of the Fort William Indian Band and wore his Ojibway raiment with great pride.

The manuscript for *Timber Wolves* was completed in 1961 but remained unpublished when J.P. Bertrand died in 1964. Draft copies, however, found their way by mysterious means into several archives where they were made accessible to researchers. Despite having been written over 35 years ago, Bertrand's work, primarily due to his close association with and in-depth knowledge of his subject, has remained exceedingly valuable. Thus, references to *Timber Wolves* have appeared in almost every study of Ontario's forest industry written since the 1960s.

The editorial committee of the Thunder Bay Historical Museum Society — Beth Boegh, Mark Chochla, Wayne Pettit, Rita Sinihelmi and Thorold Tronrud — edited the draft manuscript over the first half of 1997, eliminating some sections of the original that did not bear too closely on the story of Northwestern Ontario's timber industry, fixing a few errors of grammar and syntax, and clarifying arguments in some of the more complex chapters. Iain Hastie drew the excellent maps and Elinor Barr helped with biographical information; the committee extends its thanks to both of them as well as to Andrew Hacquoil. Thanks too to the Canadian Pulp and Paper Association for permission to reproduce graphics from their publications. The committee as a whole chose the illustrations and compiled the index. This is J.P. Bertrand's book, however, and wherever possible his words and intentions were honoured. Nevertheless, to assist the reader and to help complete the story, the committee also produced two appendixes, the first to identify further some of the people mentioned in the book, and the second to update the history of many of the mills, especially in the years since Bertrand wrote his book in 1961.

Thanks to J.P. Bertrand's daughter, Mrs. Jeanne McLean, who donated the original manuscript to the Thunder Bay Historical Museum Society and authorized its publication, we are finally pleased to make this important and entertaining work available to a wider audience.

Thorold J. Tronrud (1997)

Author's Acknowledgments

Many standard works have been consulted in the preparation of this book. This includes government reports, various Royal Commission reports, and Canada's Year Book. I wish to express my appreciation to all of those who have assisted me by their suggestions in obtaining needed information. I owe a special thanks to John Stevens, former President of the Marathon Corporation of Canada, and to Charles H. Sage, formerly a Vice-President of Spruce Falls Power and Paper Company Ltd., now Kimberly-Clark Pulp & Paper Company Ltd. I am greatly indebted to Alex Johnson for proof reading, correcting, and valuable suggestions he has made in the final draft of the manuscript.

J.P.B. (1961)

A nation with no regard for its past will have little future worth remembering. -Blake Clark

*The tree that never had to fight
For sun and sky and air and light,
That stood out in the open plain
And always got its share of rain,
Never became a forest king,
But lived and died a scrubby thing.*

*The man who never had to toil,
Who never had to win his share
Of sun and sky and flight and air,
Never became a manly man,
But lived and died as he began.*

*Good timber does not grow in ease;
The stronger wind, the tougher trees;
The farther sky, the greater length;
The more the storm, the more the strength;
By sun and cold, by rain and snows,
In tree or man, good timber grows.*

-Author Anonymous
Contributed by the late magistrate
Walter Hall Russell of Port Arthur.

Foreword

Pioneer sawmills along the Northern States, and in Canada, have been quite properly referred to by an early observer as civilizers; actually these mills formed our first North American industry. They converted our forest products into finished and useful building construction material. Their operations in Canada originally followed the Old Canoe Highway, along the Ottawa River, the Mattawa, Lake Nipissing, the French River, the north shore of Lake Huron, thence to Port Arthur and Fort William, and from there to Lac des Mille Lac, Rainy Lake, Rainy River, Kenora and Keewatin, on Lake of the Woods. This great industry, which began along the Ottawa River in the middle of the last century, had completed its cycle by 1940. All the forests, which produced suitable saw logs, were denuded of pine timber.

Towards the end of the last century, alert sawmill operators in the United States, and in Canada, began to convert some of their lumbering operations to that of manufacturing pulp and paper products, for which there was a rising demand. Two examples in the District of Thunder Bay, the operations of Alger & Smith and D.J. Arpin, need to be recorded. Their large organizations at Pigeon River included capital, which was eventually invested in Canadian pulp and paper companies. It is on this account that our first chapter begins with these international operations.

It will be seen, in reading this work, that the pulp and paper industry in Canada, and for that matter in Northwestern Ontario, developed in slow stages. It is also well known that the capital invested in these mills, although at times speculative, did not always prove profitable. The investors met with reverses, disappointments and, in a number of cases, bankruptcy. Somehow the industry always seemed to have produced leaders with vision, determination and exceptional ability. Our pulp and paper manufacturing is, as a result of such aggressive leadership, our most spectacular Canadian industry. Its mechanized woods operations, the high income earned by the specialized labour force and the high degree of comfort that was assured the bushmen in modern logging camps, would have appeared the wildest dreams to both the leaders of organized labour, and the industrialists who reorganized and refinanced this industry in the 1930s.

It has been the privilege of the author to have been closely associated with the timber industry in Northwestern Ontario since 1900, and to have occupied a good post of observation. The stories of the pioneer loggers, pulpwood operators, timber speculators and mill promoters, read like a high-class thriller. Their devices to obtain exportable pulpwood, without having to pay Crown dues, their trespassing on Crown Reserves, and their intrigues behind the scenes to gain favours with political leaders are fairly well outlined in the pages below.

It is well to keep in mind that forestry is one resource, which has been from the beginning of its development, ever under the political control or the party in power. "The government owns the forests. I have to play politics with them," remarked one of the most famous logging operators in Northwestern Ontario. The development of this industry, therefore, was so closely interwoven with politics and with politicians, that this story would have been incomplete without reference to the political leaders responsible for the administration of Ontario's provincial lands and forests.

J.P. Bertrand, 1961

Chapter One

Logging Operations at the
International Border, 1897-1920

The first sawmills ever to have been operated in Northwestern Ontario were erected along the bank of the Kaministiquia River at Fort William, and along the lakeshore in Port Arthur. The first of these mills was owned by Joseph Davidson and Adam Oliver, who had the railway contract from the Government to cut the right-of-way from West Fort William to Savanne in the mid 1870s. This mill burned down. Another one was constructed by A.H. Carpenter, which in turn was destroyed by fire. In the mid 1880s the third and most modern of these mills, since it had a bandsaw, was constructed and operated by Graham & Horne Ltd., until 1901, when it was sold to the Pigeon River Lumber Company at Port Arthur.

In Port Arthur the first sawmill was erected on the flats near the present Pool 6 Elevator. It was owned by Thomas Marks & Company. Eventually they sold it to James Conmee, a prominent railway contractor of the time. Conmee and his associates had a contract for grading a section of the Canadian Pacific Railway west of Savanne. This mill in turn burned down. Another one followed, and it was owned by Vigars Brothers, which operated until the beginning of the present century. It was dismantled when the company was merged with the interest of Captain Herbert Shear under the name of Vigars-Shear Lumber Company. They were strictly a retail organization. In the late 1920s their assets were taken over by the Thunder Bay Lumber Company headed by the late Martin J. McDonald.

Prior to the erection of these pioneer sawmills, which coincided with the first railway development, there had been no forest industry, and certainly no sawmills in this district. Except for the construction of the famous Fort Kaministiquia (Fort William) in the first decade of the last century, no cut of timber of any importance had taken place. The construction of this pretentious establishment of the Northwest Company, which consisted of sixteen large buildings, and a shipyard, entailed a major lumber operation for the times. All the lumber and material required in the construction of this historical fort was cut by pit saws.

The second operation in 1871 was strictly one of preparing squared timber for the breakwater and piers around the tiny Silver Islet mine, and like all the timber that was used in Fort William, this was cut from a pine stand nearby, since the transportation from any distance would have created a problem; moreover there were pine stands and other co-niferous forests adjacent to both operations.

White and norway pine, except for limited stands, were rather scarce along the north shore of the Lake to the present Canadian Lakehead. But here and there on the way to the boundary line, namely the Pigeon River, there were good limits of white and norway pine immediately south of the Lakehead. The first sawmills operated from these stands, which were soon cut out. By the mid 1880s, James Conmee was obtain-ing his saw logs from along the Pine River which derived its name form the extensive white and norway pine forests which covered its banks and watershed. This was considered one of the finest pine limits in existence at the time. Unfortunately, during the disastrous forest fire which consumed a vast area of timber lands all around the north shore of Lake Superior from Grand Marais, Minnesota, the Pine River timber limits were practically wiped out. Robert B. Whiteside, who held the sub-contract from the James Conmee interests, was supervising some of his operations during the summer of 1888, and nearly lost his life with one of his Indian guides on a cruising trip. Small portions of this rich limit were left standing, and taken out at a later date by Graham & Horne Ltd. of Fort William, and finally the Pigeon River Lumber Com-pany of Port Arthur in the winter of 1906-07. Some of these pine trees scattered throughout the townships of Blake, Crooks, Pearson, Pardee, Devon and along the border lakes survived this conflagration, and were taken out and cut into lumber in more recent years by individual port-able mill operators.

The largest stand of pine timber to escape the forest fire of the 1880s was along the Pigeon River and its tributaries, the Arrow and the White-fish Rivers on the Canadian side, and the Stump and the Swamp Rivers in Minnesota. Yet, the fine timber limits of the Arrow River bore evi-dence of this great fire as late as 1900. For a good half-mile, partly burned tree trunks were the stark and silent witnesses of this great forest conflagration. The pine lands in and about the watershed of the Pigeon River had been acquired in the early 1890s by two groups of loggers. One was the Alger & Smith Company of Detroit, Michigan, who at that time had purchased vast timber limits in Northern Minnesota which they were logging largely by rail haul. The other, much larger section of these timber stands, was owned by D.J. Arpin, a pioneer lumber manufacturer from Wisconsin Rapids, Wisconsin; a well-known logging personality in Minnesota, and subsequently at the Canadian Lakehead.

Alger & Smith

The Alger & Smith Company began to cut out their timber limits along Pigeon River in 1897. This was the first major logging operation at the international boundary. The general superintendent was Jack Murray, nicknamed "Sand-bar" Murray, a native of the county of Glengarry. He had emigrated to the United States with his parents as a young man, and had been associated for many years with Alger & Smith. Jack Murray was a real logger and a good river man. Possessed of an impressive physique, he commanded the respect, as well as the affection of his men. Although a hard driver when the need presented itself, he could get the most out of a crew of river drivers, particularly when engaged in breaking up a log jam, which was always dangerous work. But one thing was certain, Jack Murray himself would always be at the point of danger. A brother of his, Tom, who, like Jack, became well known at the Lakehead at a later date, was one of the camp foremen. The clerk in charge was Fred Lander. He was a sort of a comptroller over the operation, looking after its financial end.

The Pigeon River is a wild stream dropping 914 feet in something like 25 miles. In order to drive saw logs down such a precipitous river, considerable improvement had to be made to prevent some of the smaller logs from breaking into pieces. In charge of that operation was a well-known dam builder by the name of Abe Flatt. The Alger & Smith Company created two chartered subsidiaries to adjust the tolls, and to regulate the flow of the stream by building dams and flumes. One large flume was constructed at High Falls, which has a drop of 110 feet, near the mouth of the Pigeon River. These two subsidiary companies were international in scope. The first was known as the Pigeon River & Tributaries Slide and Boom Company; the second, the Arrow River and Tributaries Slide and Boom Company. According to the River Act, these two companies were to act fairly with other operators along the rivers where they extracted timber.

At the mouth of the Pigeon River, a double boom was stretched from the Ontario to the Minnesota shore to store the logs that had been

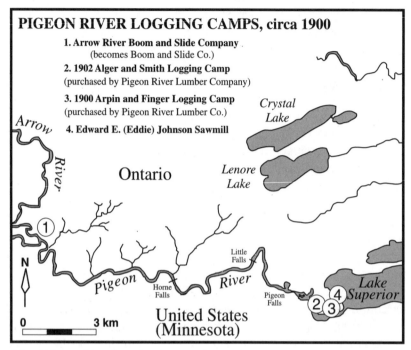

Map 1 Pigeon River Logging Camps, circa 1900

taken out during the previous winter, until such time as they would be towed out. Alger & Smith's powerful tug towed those logs to their sawmill in Duluth, which at that time was the largest sawmill in the United States, with a yearly production of 108 million feet of lumber. The Alger & Smith Company established their headquarter's camp on the Canadian bank of the Pigeon River, about a quarter of a mile from Pigeon Bay. This consisted of a warehouse, office, stables, blacksmith shop, harness shop (with Alex Hunt as harness-maker), cook camp, bunk house (for some of the men who were going in or out of the logging operation up river); and two well-built cabins with fairly modern conveniences for the superintendent, Jack Murray and his family, and Fred Lander and his family. This extensive operation of Alger & Smith, which began in 1897 along the Pigeon River, lasted until 1902, when their water rights, remaining timber stands and equipment were purchased by the Pigeon River Lumber Company. Their other holdings in Minnesota were taken over by the Weyerhaeuser Corporation.

The Alger & Smith Company, and their head, General Russell Alexander Alger, were prominent in the American and Canadian forestry industry, and in the mining industry of Northwestern Ontario. In the mid 1880s, Alger purchased the Beaver Mine from Oliver Daunais and his associates. The General and R.J. Peters, another prominent lumberman from Manistee, Michigan, operated this historic mine for a few years.

General Alger was a commanding figure in American history. He was a native of Ohio, and a graduate in Law of the State University. He served with distinction during the war between the states. When the hostilities were over he established a law practice at Saginaw, Michigan. Eventually he moved to Detroit, and became a partner in the firm of Moore, Alger & Company, dealers in pine lands and lumber. The partnership was dissolved, and another organized by the name of Alger, Smith and Company, of which he became president. At that time, Michigan was experiencing unusual activity in logging and lumber manufacturing. In 1884, General Alger was elected governor of his state, a position which he held at the time of his ownership of the Beaver Mine. In 1896 he was appointed Secretary of Defense in the newly-formed McKinley administration. He died in Washington, D. C., on January 24, 1907.

General Alger, who had had a vast experience in the exploitation of forest timber from the northern states, was alarmed at the rate that this natural resource was being denuded. Large areas of unproductive and sub-marginal soil would be left as a result of the wasteful logging methods then in use, which were permitted by the various state governments. Despairing, despite his great influence as a leading statesman and lumber manufacturer, of ever creating a climate in which the various states would be able to introduce a more orderly method of logging and a sound policy of conservation and sustained yield, he turned his eyes toward Canada.

In 1897, with his friend Sir William Van Horne of Canadian Pacific Railway fame, and a fellow American, he organized the Laurentide Pulp & Paper Company with headquarters in Montreal. They purchased a small mill at Grand Mere, Quebec, along the bank of the St. Maurice River, which included hydro power rights. They obtained a concession from the Crown lands of the Province of 1,500 square miles of lands bearing spruce of the highest quality for the manufacture of wood pulp. A sustained yield program of cutting and conservation was immediately put into effect. In time this became a model for all the pulp and paper mill operators in the province of Quebec. Another company was organized by these two industrialists, and included a third partner, R.B. Angus, director of the Canadian Pacific Railway Company. They had secured power rights on Grand Falls, New Brunswick. This particular venture, however, seems never to have passed beyond the embryo stage.

By 1902, the Laurentide Pulp & Paper Company was experiencing some financial difficulties. A disastrous fire had occurred at the plant, and unfortunately neither General Alger nor Sir William Van Horne was able to devote much time to the enterprise. Eventually new methods of mill operation were introduced into the mill. American engineers were also brought in and new capital added so that, within two years, this company had regained a commanding position in the newsprint market.

What General Alger had feared most was already happening in the United States. The wasteful logging operations carried on for some years without a proper plan of conservation and re-forestation were having a telling effect. The U.S. pulpwood resources were being rapidly exhausted, and American paper manufacturers were beginning to make some inroads into the Canadian forests. Two of these companies added many thousand square miles to their already large holdings, while still other American firms were doing the same. Even Wisconsin mills were transporting pulpwood logs from Quebec, yet only one American company was building a mill to manufacture wood pulp and paper products in Canada.

Alger, Van Horne and their partners became foremost exponents of Canadian forest conservation. In a memorial to the government of the day, they suggested the levying of an export duty on pulpwood, or the prohibition of the exportation of pulpwood to the United States altogether. "Stumps in the ground," stated Van Horne,

> are the only things we have to show for our export. One cord of
> pulpwood exported from Canada yielded to Canada an all-out
> interest less than $6.00; the same cord of pulpwood manufactured
> into paper yielded $36.00. No sane individual would waste his
> raw materials in such a way when he could do so much better with
> them, and I can see no good reason why a Government should do
> so any more than an individual.

Therefore, General Russell Alger can qualify not only as a pioneer logging operator in Northwestern Ontario, but as a pioneer newsprint manufacturer in the province of Quebec. He also deserves credit, along with Van Horne, for having introduced scientific forest management in the operation of their timber limits along the St. Maurice River. (A plaque, or a cairn, should therefore be erected to his memory for his contribution to scientific forestry management.)

Pigeon River Lumber

References have already been made to D.J. Arpin of Wisconsin Rapids. He had acquired various parcels of timber lands, both from the Crown and from private individuals, along the banks of the Pigeon River and its tributaries, on both sides of the International border. To exploit these pine limits, Arpin organized a company in 1899 under the law of Wisconsin with a paid-up capital of $300,000 which was subsequently increased to $500,000. The partners of this pioneering firm were: D.J. Arpin, Wisconsin Rapids, president; F.P. Hixon, a banker of LaCrosse, Wisconsin, vice-president; Herman Finger, a logger and lumber manufacturer of Eagle River, Wisconsin, general manager; and William Scott, insurance and land agent of Wisconsin Rapids, as secretary-treasurer. Herman Finger and William Scott paid a visit to Pigeon River in the summer of 1899 to view some of the timber which was going to be a part of the assets of the Company. No member of the partnerships however, visited the Canadian Lakehead until the autumn of 1900.

Herman Finger again visited the Pigeon River, this time with Douglas Wark with whom Finger had been associated for many years in Wisconsin. They traveled through Berth A, the timber limit along the left bank of the Arrow River, a Pigeon River tributary, before they blazed their way to Silver Mountain along the Port Arthur, Duluth and Western Railway (P.A.D. & W.). Both Finger and Wark arrived in Port Arthur in late September of 1900, and set out immediately to organize a complete logging operation.

At that time, John McCuaig, a timber contractor in Port Arthur, had been taking out piling timber and railway ties for McKenzie & Mann, who were constructing the Canadian Northern Railway line from Port Arthur to Winnipeg, subsequently the present Canadian National Railway. Finger bought all of McCuaig's equipment, wagons, horses, tents, kitchen utensils, as well as blankets. This equipment was transported to the mouth of the river, and a landing camp was erected a short distance below that of Alger and Smith's. Matthew (Matt) Tracey, an Ottawa Valley logger, was appointed to be in charge of that operation.

Temporary docks were erected, and sheds, an office, barn, bunk-house, cook camp, and two root houses went up as by magic. Subsequently Matt Tracey was appointed camp foreman.

Having laid out the plan for his operation with two logging camps under the supervision of Doug Wark, Finger left for the east on the 5th of October, 1900. While in Collingwood he contracted on behalf of his company, and that of Alger and Smith, for all the food supplies for the two organizations from a wholesale firm, Thomas Long and Brothers Ltd., who operated a shipyard, and a line of steamers. He made an arrangement with them to ship all his supplies by their boats directly to Pigeon Bay. From there, Finger moved on to Sarnia to purchase his horses, sleigh-runners, harnesses, hay and oats and other supplies which were to be shipped directly from Sarnia to Pigeon Bay. This bay, for the month of October, became a harbour of importance, with one or two steamers coming in every week with provisions and equipment. Key were the "Majestic" and the "City of Collingwood" operating out of Collingwood, and the "Monarch" and the "United Empire" sailing out of Sarnia to the Lakehead and Duluth. One small shipment was made from Duluth by the pioneer steamer "Hiram R. Dixon." One tug, the "Siskiwit," mastered by Captain Nicholas Marin, and owned by the North Shore Timber Company, of which the late James Whalen was the manager, was hired out to the Pigeon River Lumber Company from the 1st of October until the end of the season, with two scows to transfer the supplies, equipment and horses from the steamers to the shore dock. On one occasion when the "United Empire" brought a heavy load of hay and oats, including thirty teams of horses, another tug, "The Mary Ann of Dunnville," first on the register of Canadian shipping, and two scows operated by Graham & Horne, Fort William, were hired for that week-end to assist in transferring the cargo from ship to shore.

In January, 1901, Arpin's Pigeon River Lumber Company bought out the entire assets of the Graham & Horne Company Ltd. of Fort William, Ontario. This included their saw and planing mill on the left bank of the Kaministiquia, where the Murphy dock later stood, the timber cut of that winter on the Jarvis River, and a subsidiary of theirs, the Lake Superior Tug Company, which operated the tugs "Mary Ann," "Atkins" and "Salty Jack," as well as all their booms. A new tug which was under construction in the Thomas Long & Brothers shipyards at Collingwood, under a contract with Lake Superior Tug Company, was transferred to its new owners. It was named "Laura Grace" after Mrs.

Herman Finger's maiden name. This fine tug, mastered by Captain Nicholas Marin, remained the pride of the Lakehead harbours until the fall of 1905 when the icebreaking tug "James Whalen" was brought to Port Arthur.

During the season of 1901, the old Graham & Horne mill at Fort William was operated by its new owners on a makeshift basis. The Pigeon River Lumber Company began the construction of a large saw-mill in the early summer of that year in Port Arthur on property leased from the Canadian Pacific Railway near the old King's Elevator (later Manitoba Pool 2) between Van Horne and McVicar Street. They built a large planing mill west of the Canadian Pacific Railway track be-tween McIntyre and Stephen Streets.

This sawmill, one of the most modern of its kind, began operations May 21, 1902. Its equipment consisted of two band saws, one horizon-tal resaw and two gang-edgers. It averaged 200,000 feet of lumber every 20 hours. The planing mill was equipped with the latest machin-ery, with sufficient capacity to process lumber in proportion to the mill run. There was a good local market for their lumber, but by far the greatest portion of the cut was shipped to Manitoba which, at that time, was experiencing a building boom. The Pigeon River enterprise, both in town and in the woods, was the pay cheque for many a Port Arthur family. It was the first large industry in the city until the construction of the Port Arthur Shipbuilding Company in 1911.

The Pigeon River Lumber Company's Gunflint timber limit was located in Minnesota over the height-of-land. Thus it became necessary to log it by rail, since logs could not be floated into Lake Superior. The company obtained a Charter, and constructed a spur twelve miles in length on the American side of the border to connect their limit with the old P.A.D. & W. Railway at North Lake. This road was known as the Lake Superior and Gunflint Railway. Its motive power consisted of two small locomotives, one of a Lima type. The logging operation of the Gunflint limit was under the supervision of Oliver Barton, a well-known Minnesota logger. All the equipment and supplies had to be brought in to Port Arthur from Duluth "in bond," and shipped to Gunflint via the P.A.D. & W. and the company's own railway. The logs were sent by rail to Port Arthur, and shunted to the company's spur alongside the hot pond to be unloaded. For a few years Port Arthur had a substantial winter sawmill operation. With their Gunflint limits cut out, the rails were taken up, and the two locomotives sold. One of these was operated

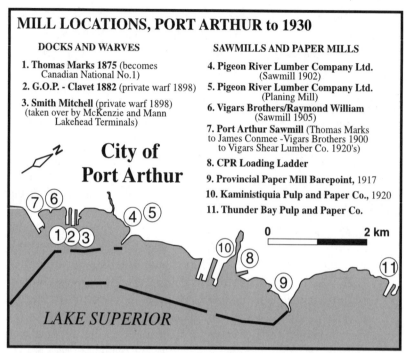

MILL LOCATIONS, PORT ARTHUR to 1930

DOCKS AND WARVES

1. **Thomas Marks 1875** (becomes Canadian National No.1)
2. **G.O.P. - Clavet 1882** (private warf 1898)
3. **Smith Mitchell** (private warf 1898) (taken over by McKenzie and Mann Lakehead Terminals)

SAWMILLS AND PAPER MILLS

4. **Pigeon River Lumber Company Ltd.** (Sawmill 1902)
5. **Pigeon River Lumber Company Ltd.** (Planing Mill)
6. **Vigars Brothers/Raymond William** (Sawmill 1905)
7. **Port Arthur Sawmill** (Thomas Marks to James Conmee -Vigars Brothers 1900 to Vigars Shear Lumber Co. 1920's)
8. **CPR Loading Ladder**
9. **Provincial Paper Mill Barepoint,** 1917
10. **Kaministiquia Pulp and Paper Co.,** 1920
11. **Thunder Bay Pulp and Paper Co.**

City of
Port Arthur

0   2 km

LAKE SUPERIOR

Map 2 Mill Locations, Port Arthur to 1930

for many years at Hewitson's quarry.

Such were the conditions in transportation when the Pigeon River Lumber Company entered the field as active operators. It has already been stated that the Pigeon River was a rough waterway, needing considerable improvement to make possible the driving of saw logs — likewise toting supplies to the various camps became a rather expensive task. Many of the smaller logs were broken in two in driving them down to the mouth of the river, and much of the pine timber was over mature. The retail market did not command a very high price for lumber, even though the demand for it was consistent. There was an off-grade run of lumber called No. 6, of which there was a fairly large quantity in 1903 and 1904. This was selling at $5.00 a load to the farmers who were taking up lands in the townships of Oliver and McIntyre. This lumber can still be seen on the barns and other buildings erected on farms at the time. The six- and eight-foot lengths of lumber, a large quantity of which was inevitable due to considerable log breakage, sold at $2.00 per 1,000 feet below the regular prices.

The situation must have been only too evident to a well-trained lumberman such as Herman Finger, and may have been a factor in making him decide to resign from the management of the company in the summer of 1904. In the autumn of that year he made a trip to what is now Northern Manitoba to look over the spruce timber forests of the northwest. He returned much impressed with what he saw. From 1905 until 1912, with his faithful timber cruiser, Douglas Wark, and his son, Oscar, he explored, cruised, estimated and acquired vast timber rights, particularly along the watershed of the Carrot River, a tributary of the Saskatchewan River. From the beginning of his exploration he made his headquarters along the Saskatchewan River at The Pas, a trading post and mission then in the Northwest Territories, and now in the province of Manitoba. By 1913 he formed a partnership with C.R. Smith, a prominent industrialist of Oshkosh, Wisconsin, and a modern mill was erected which began operation the following year. This firm was known as Finger and Smith Ltd., and operated successfully until the fall of 1920. Herman Finger, who had never admitted to himself that relaxation was necessary, began to realize that he was aging. He and Smith, then sold their entire interests to the Prince Albert Lumber Company Ltd. This firm immediately organized a new company, The Pas Lumber Company Ltd. It was a successful operation until 1957, when their timber limits were cut out. They then folded up. It was the usual tragic logging story — cut out and get out. Finger had moved to Winnipeg, and lived in retirement until his death during the winter of 1928-29.

Herman Finger possessed a picturesque personality. He was a rugged, and a most original man. A native of Milwaukee, Wisconsin, at eighteen years of age he was already in charge of logging operations and river drives. As already stated in a previous paragraph, he pushed ahead until he became the owner of the Eagle River Lumber Company at Eagle River, Wisconsin. Before moving into Canada, he owned and operated a large farm in his own state, and did a great deal of timber cruising and profitable timber speculating. When he came to the Canadian Lakehead in late September, 1900, he created quite a sensation. He was a huge bulk of a man, weighing 260 pounds. He had a ruddy complexion, a reddish moustache and a Vandyke beard. Many are the stories that circulated about Herman Finger. Amazingly agile and supple for a man of his size, Finger could walk miles without tiring, and was much at home on floating saw logs, even on fast water. There was not a thing about woods operation with which he was not familiar. He was handy

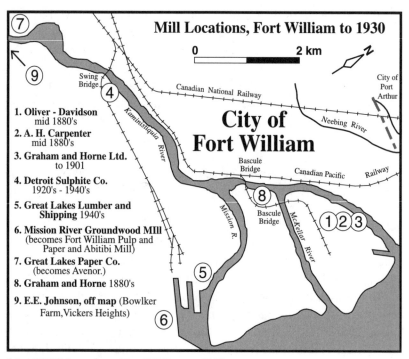

Map 3 Mill Locations, Fort William to 1930

with tools of any kind, and possessed good judgement. He got along well with his men. He was fond of horses, and purchased many fine-looking animals for his logging operations. He would not tolerate a whip in the hands of his teamsters, and would have dismissed anyone who disobeyed his orders. He possessed an unusually good voice, and expressed himself rather slowly; his words were like sounds from a bell. While he could be choleric on occasion when things went wrong, he was never heard to use profane language. He was not afraid of man or beast. On one occasion in opening up the drive along the upper stretches of the Arrow River, he more than held his own with a peavey in his hands with Big Jack "Sand-Bar" Murray, with whom he had had an argument as to whose logs should be floated first down the river.

The best story ever told of Herman Finger was on the occasion of a visit to the Canadian Lakehead by the Duke and Duchess of Sutherland in the summer of 1903. They were traveling by Canadian Pacific Railway private car to Alberta where the Duke had purchased a large tract of land for settlement. While in Port Arthur their car had been switched to a spur near the Pigeon River Mill. From there the Duke paid a visit to Joseph G. King, a councilman of the Town of Port Arthur, who operated, under lease, the first elevator to have been erected at the Lakehead, and known as "King's Elevator," later Manitoba Pool No. 2. This elevator was known also as a "hospital elevator," and it was on this account that the Duke of Sutherland was interested in visiting it. King, who was a native of England, had received a good education, and was the last word in proper deportment in the presence of royal personages. With a good background and rather aristocratic bearing, he was right at home with the representatives of the higher stratum of British society. After the usual preliminary introduction, the Duke of Sutherland visited this unique grain-treating elevator. Eventually he manifested a desire to see the large sawmill operated by the Pigeon River Lumber Company. J.G. King, who knew Mr. Finger well, brought the Duke over to see him at the office, and told the clerk at the front desk that he wished Mr. Finger to meet a distinguished visitor from Britain. Mr. Finger immediately invited them in. Finger had never learned the art of dictating a letter to a stenographer. He always wrote his messages and communications longhand, and he was busy writing a letter to his logging superintendent, Douglas Wark. J.G. King, with hat in hand, walked in with the Duke, and as he faced Herman Finger, he said, "Mr. Finger, I wish to introduce you to His Grace, the Duke of Sutherland." Said Mr. Finger, "Pleased to meet you Duke — sit down — I'm busy now writing a letter to Doug. I'll be with you in a minute." Finger was a blunt man with a heart of gold. After visiting the sawmill with Herman Finger, so impressed was the Duke with this rather original character, the like of whom he had never met, that he invited him to dinner on board his private car. It was reported that he had a wonderful time. Mr. Finger was much impressed with the refreshment, the food, and the Duke's wife, as he referred to the Duchess. Finger was a real pioneer lumber operator, and left his mark deeply imprinted on the memory of all the men who worked for him. In his permanent residence at The Pas, he was the mayor of the town for many years and, at times, police magistrate. He fitted in well at the frontier.

William Scott, then secretary-treasurer of the Pigeon River Lumber Company, with headquarters at Wisconsin Rapids, Wisconsin, moved to Port Arthur with his family early in 1904, and was appointed general manager. George S. Clarke of Minneapolis, an experienced mill operator, who had been associated with the Carpenter interests for some years, was appointed assistant general manager. In the autumn of 1910 he resigned to accept the position of manager of the Tremont Lumber Company of Winfield, Louisiana. Later he became a partner, and was associated with the management of the Eastern Coast Lumber Company. He died in Minneapolis in the summer of 1945. "Tarpaper" Peter Grant was appointed logging superintendent, to succeed Douglas Wark, and remained with the Pigeon River Lumber Company to the end of their operations. Bill Cochrane was a timber cruiser with the same firm. By 1920, with their timber limits cut out, the Pigeon River Lumber Company folded. Their large mill was dismantled in 1928, and the machinery was sold to the Burrows Lumber Company of Winnipeg, who were operating a mill in Northern Manitoba. A number of businessmen engaged in the lumber business had their beginnings with the Pigeon River Lumber Company. The late D.J. McDonald of Winnipeg, J.A. "Sandy" McDonald (Vancouver), George McDonald (Cleveland), Clayte Sherry (Saskatoon), Martin J. McDonald (Port Arthur), and J.H. McLennan (also of Port Arthur), Angus G. McCormack (president of the City Feed Company), and Clarence H. Moors (president of the Mount McKay Feed Company), also had their early training with the Pigeon River Lumber Company, as did George Matthews of Fort William, to name only a few.

It is doubtful if the Pigeon River Lumber Company, despite aggressive and careful management, ever broke even on their original investment. It was the story of many pioneer sawmill operators of the time. The quarter interest share which Herman Finger originally held in the company was purchased in the early autumn of 1904 by Alexander Stewart, president of the Alexander Stewart Lumber Company of Wausau, Wisconsin. He in turn divided these shares in two, keeping one-half for himself, and turning over the other to Walter Alexander, a partner in his company. Both Alexander Stewart and Walter Alexander attended a meeting of the shareholders and directors of the Pigeon River Lumber Company, which was held in Port Arthur in July of 1906. Alexander Stewart, already aged, died on May 24, 1912.

By the late 1890s, the vast forest lands of Wisconsin had been practically denuded of red and white pine. This great wealth was dissipated through shortsightedness and lack of a progressive forest policy by the government of that state. This, as has already been seen, had prompted Alger & Smith and Company, and D.J. Arpin, the founder of the Pigeon River Lumber Company, to purchase timber rights along the banks and watershed of the Pigeon River. It must have been the same motive that prompted Alexander Stewart to purchase Finger's quarter interest in the Pigeon River Lumber Company.

Operating as Stewart's partner in Northern Wisconsin was Cyrus C. Yawkey, an intensely progressive and imaginative industrialist. He conceived the idea of utilizing what was left of the forest — spruce, balsam, jackpine and poplar — in manufacturing wood pulp products. Shortly after investing some of their capital in the Pigeon River Lumber Company, Stewart and Yawkey organized a new company, of which Yawkey became president. A pulp and paper mill was erected at Rothschild, Wisconsin, at a cost of one million dollars. It was reported at the time that "Alexander Stewart was sweating blood" — he had never heard, nor had he ever experienced such a costly undertaking in his life. He feared for a while that he would go broke. He told his friend, D.J. Arpin, that he was afraid this mill would cost two million dollars. This was the origin of the Marathon Paper Company, later the Marathon Corporation of Canada Ltd., of Marathon, Ontario.

This corporation is no stranger to Northwestern Ontario. Some of the capital and members of their original executives have been active in manufacturing lumber at the Canadian Lakehead, as partners in the Pigeon River Lumber Company, from 1904 to 1920. During 1920, the partners of the Pigeon River Lumber Company, with their pine limits cut out, decided to wind up the affairs of the firm, and for that task they appointed a young lawyer from Wausau, Wisconsin, Edward E. Johnson, a man destined to play a prominent role in Northwestern Ontario. Ben Alexander, a son of Walter Alexander, became a partner in the E.E. Johnson enterprises. There was, therefore, continuity of capital investment in Northwestern Ontario from the same groups that had purchased the Finger interests in the Pigeon River Lumber Company in 1904.

Chapter Two

The Men of the Bush

T his book would be incomplete without some reference to the
bushmen or lumberjacks, the equipment they used, and the types
of camps that were constructed during that period of history.
Logging camp construction had evolved little in the nineteenth century.
Timbers were laid one upon another, and were joined at the end in a
dovetail lock. The larger the timber, the more economically the camp
could be built. The cracks were chinked with moss, and covered with a
clay mortar. The roofs were covered with the traditional tar felt laid
over poles. The floors were made of hewn logs, and one or two vents
completed the whole of the structure with possibly a sash at each end of
the camp. Poles were used in the construction of men's bunks, double
deck style.

By the late 1890s it became imperative to build better logging camps
in order to comply with the new health regulations of the northern states,
and of our own province of Ontario. Bunkhouse walls were made higher
with more roof ventilation, and more lights. Lumber was used for the
floor, in making up the bunks for the men, and as roof board which was
covered, as formerly, with tar felt. At one end of the sleeping camp was
a wash stand where men could wash themselves before going to break-
fast or upon returning from work in the evening. This short distance
(some fifteen feet) between the door of the sleeping camp, and that which
opened onto the cook camp, was also covered so that men could go from
one camp to the other without being affected by a heavy downpour of
rain or snow. Otherwise camp architecture remained pretty much un-
changed.

In the centre of the bunk house was a large box stove, the heat from
which kept the camp comfortable during the cold winter months. One
of these sleeping camps, in a major operation, would house from 90 to
100 men. After working hard in the deep snow all day, the first thing the
men would do on arriving at the camp was to remove their rubbers or
moccasins and socks and, in many cases, wet trousers. All these would
be hung on wires which had been strung above the stove so that the men
would have dry socks, moccasins and trousers to wear the following
morning. Since these men rarely had occasion to wash their feet, and

perspired a great deal in their daily toil, it can well be imagined that the odour emanating from these hundreds of pairs of socks was rather obnoxious to anyone possessing a delicate olfactory organ. From supper until 9 o'clock when the lights were turned out, and everybody had to be in his bunk, the men were busy at various tasks. Some mended their clothes, others sharpened their axes or the points of their cant-hooks, and others smoked or just gossiped. A few of them, particularly among the teamsters and loaders, used to find their way into the stable for a friendly game of poker, using a bale of hay as a table, and squatting on the stable floor. The stakes were usually plug chewing or smoking tobacco.

These vigorous bushmen, hard workers all of them, were endowed with ravenous appetites. The food was particularly good, even for those days, although they did not have all the trimmings of today. The fare was substantial, with plenty of pastry. Companies procured the best of meat and produce, and hired competent cooks to prepare the food. These cooks received the same salary as a camp foreman! The cook camp was kept clean, and the walls pretty well decorated with American election posters of various kinds. The battery of stoves were the best that could be obtained from manufacturers. The cost of feeding a man per day in one of these camps was not high, averaging about 35¢ to 40¢.

Once they retired, two men occupied a bunk, either the lower or the upper part. The mattress consisted of a layer of hay, with a blanket for a sheet, and two blankets for a cover. Then the symphony began. Once the camp had been in darkness, it was an amusing experience for the sensitive ear to listen to the various noises emitted from this sleeping horde of woodsmen. Some would be yawning, others would be snoring, and a number complained of old pains. Considering that the fare of the dinner consisted largely of well-baked pork and beans, few sounds from the various instruments of a pipe band were missing. This has been referred to by Stewart Holbrock, the well-known American author on forest lore, as a "nocturne."

The staff of a major logging camp in the late 1890s and early years of the twentieth century, consisted of the foreman (the Push), the "bullcook," who looked after the fuel and the water for the camps, and kept the fire going, the "straw push," who had charge of the main road, and the clerk. In some cases the clerk was the scaler as well, thereby representing both the interests of the organization he worked for and the Department of Crown Lands. The blacksmith had his own shop which was shared by the saw filer. Both were very important mechanics in the operation of a logging camp. There was also a small shop for a handyman, a gifted craftsman who could perform miracles in repairing broken logging equipment, whether it be a whiffle tree, tongue, bunk, wagon, or sleigh runner. He was usually able to obtain his material by going down along the river bank, and picking up a hardwood tree, mainly ash, which he then cut to the needed length. The main road teamsters would bring it up to the shop on their return trip. All in all, the half a dozen buildings which made the whole of the logging camps, formed an interesting little village in the wilderness.

The office housed the camp foreman, the clerk, and at times the logging superintendent, along with one or two officials of the firm when visiting the operation. The interior was rough but neat. There was a high desk for the clerk, and shelves for the camp books which had to be posted regularly. Behind it, along the wall, were several shelves containing work clothing, mitts, socks, moccasins, boots, stamps, writing pads, envelopes and medicines which were sold to the bushmen. It was also the post office for outgoing and incoming mail. In the logging vernacular it was the "van."

With nothing to do on Sunday evenings, the Lord's Day, and crowded together in idleness, some bush men lost their patience and their tempers. But, except for an occasional brawl, the men seemed to get along pretty well together. Unfortunately camp life was marred by lack of proper regulations for cleanliness. A laundry, even of the crudest kind, was unheard of. A few of the woodsmen would have enough ambition to get up on Sunday morning, and wash their underclothes and socks, walking down to the creek nearby, and usually warming the water in a kettle left there for that purpose. Some of them boiled their clothing. Generally speaking, not one in five of these bushmen, at that particular period, ever washed their underwear from the fall until they discarded it in the spring. Not all of them stayed for the whole winter. There was always a transient group of bushmen who came and went from camp to camp, and from company to company some of them remaining for a few weeks, others for only a few days. Most of these "bushwhackers" carried a considerable number of parasitic insects about them. In short, there was a constant immigration and emigration of lice to and from logging camps. As one lumberjack of the time quaintly stated it, "I've brought my lice from as far east as the Ottawa valley," although he had been in the district for two years.

Saturday evening was, by custom, the social event of the week. The men could stay up until 12 or even 1 o'clock a.m., and the time was spent in singing, card playing, story telling and dancing. There were usually some pretty good step dancers among these bushmen, but square dancing was the popular type of amusement. They were divided into two groups, male and female, with the men who took the part of the girls having a handkerchief, necktie or anything around the left arm to set them apart from the men. Then they would have a real hoedown. There were always musical instruments at the logging camps; sometimes accordions and mouth-organs, but usually violins. A veteran bush man, by the name of Hutchison, who looked after a part of the main road, could play this instrument. There were few tunes that he had mastered, but one of them was "Pop Goes The Weasel." He used to sit and play on top of his bunk and, with the manipulation of his gnarled fingers on the strings of the old fiddle, he could literally make the weasel pop.

With the coming of spring, when it was impossible to maintain ice roads, they would break up the camps. The bushmen would then throw off all their old clothes on a pile outside of the camps, which they called the "jack pot," and it was quite a pile. They would then put on the best clothes that they had to get to town. Two or three men, jack-of-all-trades type, who could cook their own meals for the next two or three weeks until it was time for the log drive on the river, would be left to look after the camp and the horses. Shortly after the men had left the camp, Native women from Grand Portage would come to gather up all the clothes from the "jack pot," and carry them away on toboggans to be made over for the members of their families. A day or two after the men left camp, they would arrive in town where they discarded their old clothes, and purchased a brand new outfit from head to toe. With the help of the local tonsorial artists, they became dandies in a matter of hours, but it was in appearance only.

Then came the roaring good times. Many of them would dissipate their hard-earned cash in a few days. The average bushman estimated that he spent approximately $25.00 a day, not counting his new spring outfit. These vigorous lumberjacks came to town in good physical condition for a "blow-out," and they seemed to be satisfied that they were receiving their money's worth. A good part of their leisure time in and about the bunk house during the winter months had been spent in planning their orgy. In their idle, profane and sensual discussions, it was not the fate of the country or social relations that they discussed, but rather the bars and bartenders they knew, and the madames who kept the bawdy houses, the inmates of which they knew by their first names. Here the men expected to relieve their lust and, incidentally, their pocket books. In the spring of 1902 an article appeared in the *Times Journal* of Fort William, entitled "The Lumberjack in Town." It related how in a matter of four or five days one lumberjack had dissipated $100 to $125 at the local bordellos. When his money ran out, the madame was heard to remark, "I wonder if he has any money in the bank." Such were the habits of the bushmen in that pioneering period of logging. Many of these men would be re-engaged, and sent back for the river drive. This was hazardous, but profitable work. They actually got $2.50 a day and board, working from daylight to dark, for three weeks or a month. They would then make another stake. For the summer months they would be employed at tasks other than logging. By fall, however, these bushmen couldn't resist the call of the tall timbers, and they would again be seeking jobs in logging camps.

During the first decade of the twentieth century, there were three madames who were operating bordellos in the intercity area between the two Lakehead towns. At that time, this area was covered by a new growth of coniferous trees which easily sheltered, and concealed these establishments from Fort William Road and the street railway track. There were, however, side roads leading to them, and everyone was aware of their existence. Livery stables of the two towns did a highly profitable business by hiring out horses and buggies, or sleighs, depending on the season of the year or the condition of the roads, to the gallants who frequented these houses of iniquity, as they were referred to by the indignant members of the clergy.

The men's clothing stores, the hotels, as well as the brothels did a rushing business, all of which was conducted on a competitive basis, according to the best tradition of our free enterprise system. One of the madames, by coincidence, bore the name of a president of the United States of that time, and the evil gossipers at the Lakehead insisted that she was his first cousin. The most imaginative and enterprising madame conceived the idea of adding some oriental flavour to her establishment. She brought in a number of Japanese inmates from the Pacific Coast and, during that one year, these women evoked considerable interest among the bushmen of the district. It was customary at the time for storekeepers to send one of their clerks along the P.A.D. & W. railway to meet the bushmen on their way to town. They would give each man a card from the store where they were employed and, without making themselves obnoxious, invite them in to buy their new outfits from them. An enterprising madame followed suit by sending one or two attractive girls to do some pandering with the bushmen on the trains. They would go up on the train one day and back the next, sometimes renewing former acquaintances.

The best known of these madames, a kindly woman in many ways, helped a great number of down-and-outers who came to her house for a meal. They were never turned away. She was the pioneer of all the bawdy house operators and outlived them all by many years. She set up her first establishment during the years of the silver boom in the mid 1880s, and even had branches at mining villages or mining camps. An old timer stated rather humorously that she had actually anticipated the gas and oil service stations by nearly half a century. Her name became familiar to everyone. According to one apparently reliable story, she became known on two continents. It was during the First World War, when many Canadians were posted on the battlefields of France and Belgium, that a Canadian regiment came to relieve a British regiment which had been on trench duty for some time. From the enemy trenches only a short distance away, came the voice of a German soldier, who had undoubtedly had been working in Northwestern Ontario, possibly on railway construction, and was well acquainted with the Canadian Lakehead. He had noticed the movement in and about the Allied trenches, and somehow made out that they were Canadians, probably from their profane language. "Hello Canadians," he called out, "which regiment are we facing now?" One of them replied, "Lake Superior, 52nd" to which he shouted back, "How's Old Mag?"

The majority of these woodsmen were bachelors, though some were widowers who could not resist the glamour of the logging camp and a return to their former type of employment. Others were grass-widowers who, not being able to get along with their mates, sought solace in the company of men, hard work, good food and an enjoyment of outdoor life. A fairly good number were decent men who worked at logging during the winter, and returned home in the spring with their earnings to add to the family income. Most of these men had traveled and worked in various logging camps from Maine to Minnesota, and from the upper reaches of the Ottawa Valley and along the Georgian Bay to the Canadian Lakehead. On the whole they were interesting types, individualists and handy with any kind of tool. They could drive saw logs on the rivers as well as they could drive horses. They could swing an axe, pull a cross-cut saw in felling a tree, or cutting it into saw logs, or handle the cant hook with the ease of a master magician.

At the dawn of the twentieth century, approximately 25% of our woodsmen were French speaking while the rest were from English-speaking areas of Eastern Canada or across the border. These two original stocks however began to disappear once the Finns made their appearance as bushworkers. Brought into this country originally to work on railway construction, but being by tradition farmers and excellent woodsmen, Finns began to gather in and about the Canadian Lakehead to take up homesteads and seek employment in logging pulpwood and railway tie camps. If rather outwardly distant, they were nevertheless home-loving and hospitable people, and a clean lot. The first thing they did on going into a camp to work was to build, at their own expense, a steam bath (sauna), something unheard of by either Canadian or American woodsmen. In this stubborn habit of theirs of keeping clean, they rendered the woodsmen of the Northern States and Northwestern Ontario a distinct service. They did not pioneer the timber operations in Northwestern Ontario, but they did pioneer cleanliness in logging camps.

The Logging Operation

Taking out a large saw log cut required considerable planning and exacting operational details. There were the elements to contend with: too much snow, extreme cold or mild weather which prevented the making of ice for the main highway, at times a shortage of labour, and a drop in lumber prices. A large amount of capital had to be tied up from the early autumn, when men, supplies and horses had to be sent into the woods, until the last log was taken out of the river, towed to the mill to be sawed and piled in the lumber yard for three or four months to dry in order to be further processed into construction materials. Sometimes many more months would pass before operators could get their returns from the many Manitoba dealers which they supplied with lumber.

The prevailing wages during the first decade of the twentieth century were $35 a month for main road teamsters, and $30 a month for skidders, that is, if they stayed all winter. Otherwise they would be paid off at $30 or $26. Usually those type of men stayed for the entire season. Sawyers and axe men received $30 a month while beginners and swampers, who were usually assigned to road cutting for the skidders, got from $26 to $28 a month. The cant hook men would receive the same salary as teamsters; that is, those who had experience, and were reckoned as first-class cant hook men, mainly the toploaders and the ground men. There used to be a beginner on the ground whose function was to line up the logs for the ground man. He was usually paid $30 a month. If he was promoted during the winter either to the job of sending up the logs or receiving them on top of the load, he was paid $35 a month. All these wages, of course, included board, but there was also $1 deductible for the hospital fee.

There were bound to be accidents and illnesses. These would include fractured limbs, particularly among cant hook men, but rarely would there be a fatal accident in felling trees. To give the men proper medical attention, a camp was erected for that purpose under contract with a Lakehead doctor. At Pigeon River it was Dr. George Wallace Brown who had the contract. One of the earliest doctors assigned to this woods' hospital was Dr. A.H. Williamson. He practiced in Port Arthur for a few years afterwards, and then moved to London, England, where he became a successful practitioner. He was succeeded, as a camp doctor, by Dr. Thomas Dow Macgillivray who was in charge of all the camps for the Pigeon River Lumber Company. He eventually established a good practice for himself in Port Arthur. His cultural contribution to Northwestern Ontario was the founding of the Macgillivray Pipe Band of Port Arthur which will ever perpetuate his memory.

The main road teamsters formed a sort of aristocracy in the logging

Fig. 1. A load is being delivered on skids over a shear skid (a) and a "floater" (b). A roller is piling wood with a "jack" (c). From Pulpwood Skidding with Horses, *WSI #694 (B-8-c) (Woodlands Sec. Cdn. Pulp & Paper Assoc., 1943), 78.*

camps. They were usually middle-aged men who had had considerable experience in handling horses. They were entrusted with capital; a good team of horses, fully harnessed and a sleigh load of logs represented a good sum of money. These teamsters were rather proud of their work. The loaders came next in the camp social strata. They were younger men with lots of experience in directing a log from the skidway to the sleigh, or receiving it on top of the load.

The logging operations consisted first of all of cruising the timber carefully, and then mapping the main roads which had to follow the down grade along a creek, since no team of horses could have pulled a large load of logs up grade. The men were set to the task of cutting these various roads, the clearing for a landing on the river banks where the logs would be unloaded on the ice, and also the various skidways along the sleigh roads. It was largely a horse and hand operation. There were two- or three-men gangs of sawyers cutting 80 to 120 logs respectively per day. The logs were piled on the skidways in the fall to be loaded on sleighs during the hauling season which began late in December and ended with the spring break-up. Some of these logs were loaded directly from skidding operations without having been piled on the rollways.

The heaviest single investment in a logging operation was the construction and maintenance of the main sleigh road from the forest to the river bank. Early in the fall, a gang of men would be employed in cutting the trees on the width of this main highway, which had already been blazed by the cruiser or the logging superintendent. Once the trees had all been cut and removed, the stumps would have to be taken out, quite a task in some cases. Then, by the use of the grub-hoes, another gang of men would level off the whole of the roadway. This road, as already stated, followed the natural declivity of a creek, for the simple reason that to pull a sleighload of logs of that weight, a down grade was essential. Moreover, the creek was dammed at different places along the roadway, and it was from these pools that they obtained the water to ice the road. By mid-December the ground along the roadway would be pretty well frozen if it was a normal winter. Icing operations would then begin and would be completed by the time hauling was in progress, a week or ten days later, depending on the weather.

The equipment used in icing the road consisted of a large water tank made up of tongue and grooved planking, and drawn by a team of horses. On top of the tank, which was absolutely water-tight, was an aperture

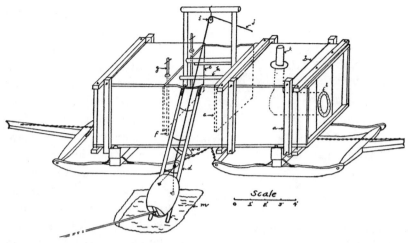

Fig. 2. A two-way, large water tank with a heater for a four-horse team. Tank measures 4' x 7' x 14' with a 390 cu. ft. capacity. (a) frame timber cross section, (b) ¾" brace rod, (c) interior partition to prevent splashing, (d) barrel ladder, (e) sheet iron trough for deflecting water to ruts, (f) holes in bottom of tank just over trough 'e', (g) plugs to open and close holes 'f', (h) a bumper to prevent barrel from falling into tank, (i) hoisting block, (j) cable, (k) stack of tank heater, (l) fuel door, (m) water hole. From Pulpwood Hauling with Horse and Sleigh, WSI #706 (B-8) *(Woodlands Section, Cdn. Pulp & Paper Assoc., 1943), 28.*

sufficiently wide to accept a barrel of water. Above the aperture was a frame about three feet high to which was attached a pulley. A sort of skid ladder consisting of two peeled skids of substantial strength were fastened together from the back, and installed when in use from the water pool onto the topside of the water tank, as a track for the water barrel. A 45-gallon barrel was used for filling the tank. In the centre of the barrel, at its bottom, was a hook to which a ring was attached. That ring, in turn, was fastened to a ferrule at the end of a hardwood pole. The decking line was fastened to the handle on top of the barrel, and would then be put through the pulley on top of the tank. A team of horses would be unhooked from the tongue, and the main double tree hitched to the decking line. The teamster would drive his team while his helper would be at the end of a long pole directing the operations. His function was to lift the bottom of the barrel as soon as it dropped down to the surface of the water and fill it, then give the signal to the teamster to go ahead and, as the barrel reached the edge of the opening on top of

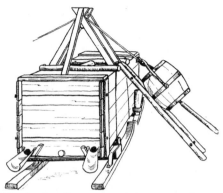

Fig. 3. Smaller water tank commonly used in Quebec. From Pulpwood Hauling with Horse and Sleigh, WSI #706 (B-8) *(Woodlands Section, Cdn. Pulp & Paper Assoc., 1943), 29.*

the tank, it emptied itself. The helper would then shout "back down," and the team would be backed up to the tank, and the operation would be repeated again and again until the tank was filled. This operation was always done at night with torchlights. In 20- or 25- below-zero weather it was a task to perform worthy of heroes. Then the helper would stand on a platform at the rear of the full tank and direct two jets which sprayed water along the roadway at a proper depth depending on the speed of the team. This would require many nights of work until the road was completely iced.

After the icing had been completed to the satisfaction of the camp foreman, or the logging superintendent, it was necessary to groove or rut it. For that purpose D.J. Arpin, the Wisconsin lumberman who founded the Pigeon River Lumber Company, had invented and patented a special device which consisted of two large steel chisels of a semi-oval design. These chisels were inserted on the sides of two heavy runners made of norway pine fastened together by means of bunks and braces, and then pulled by a team of horses to carve the grooves. Then a man would be sent along to clean out the grooves. Once the sleigh load of logs got into these grooves they could not get out until they reached the landing. It was the equivalent of a grooved railway track. It had great advantages inasmuch as it permitted the hauling of very heavy loads with no chances of sliding off the highway. On the other hand, a return-ing teamster, with an empty sleigh, had to drive off to the side in a hurry when meeting a loaded sleigh. Each teamster would yell out his particu-lar signal which could be heard at quite a distance. Even then the odd collision would happen, and a heavy load of logs, team and all, would pile onto an incoming team with an empty sleigh. That was a wreck to clean up.

From the main road there would be a number of side roads leading to skidding operations or to rollways where logs had been piled up in the fall. These roads were usually down grade. If it wasn't too steep, hay was used to control the speed of the sleighs. On steep grades, however, sand, heated in a number of nearby fires, was used to limit speeds. As the team was coming down, one man, who had one section of the road to look after, would drop some hot sand from a pail just ahead of the runners as the sleigh was passing along. Sometimes this would stop the sleigh altogether, but as soon as the sand cooled off, the gravity was sufficient to get the sleigh moving by a command of the teamster to his horses to "get on." The men entrusted with the maintenance of these side roads, whether they were dropping hay or sand, were referred to as "road monkeys."

There has been a reference made to a "decking line" which was used to fill the water tank from the pools along the creeks. Actually this decking line was one of the most important pieces of equipment. It was used to build up the rollways during the fall, and to load the sleighs with the logs during the hauling winter period. The decking line was light but strong, forged from tested steel. A breakage during its use would have resulted in serious accidents.

The loading of saw logs on the sleigh was as follows: with the sleigh placed close to the skidway or rollway, two stout skids with sharp prongs at the upper end were placed from the rollway to the end of the sleigh bunks. These bunks were held to the sleigh beams by one large steel pin which gave them a free swivelling action. The first two logs to be sent up to the sleigh would be fastened securely at the end of each of the two bunks by means of chain corner-binds.

Fig. 4. End loading of 4' wood. Teamster faces the end of the pile and advances the sleigh forward as necessary. A "packsack" is attached to help compress the snow and smooth out the road. From Pulpwood Hauling with Horse and Sleigh, WSI #706 (B-8) (Woodlands Section, Cdn. Pulp & Paper Assoc., 1943), 106.

This was done by the use of a fit-hook which could be easily disengaged by a blow from the bottom end of a cant hook, when the load was being unloaded at the river landing.

These sleigh bunks, according to the regulations of the time, were 14' in length, and could therefore carry a very large footage of saw logs. Once the first two logs were made fast to the ends of the bunks, the skids were then applied to the side of the first log, and loading proceeded rapidly. The loading crew consisted of three cant hook men, a lead man with one helper on the ground, and a top man on the sleigh, referred to as a "top loader." After the space between the two logs on the bunks had been filled, another log would be placed on each side on top of the first row, and the space filled with other logs, then likewise a third row would begin when a set of wrappers would be put on around the load and fastened on the unloading side by a fit-hook, which could be disengaged in the same manner as already referred to above. Sometimes another row of logs would be added, and this would necessitate another set of wrappers. To bind the load together, two logs, one on each side, would complete the load, and on top of the wrappers two heavy logs would be placed, the weight of which would act as binders, and make the sleigh load of logs secure until it reached the landing or the "dump" as it was sometimes called.

The motive power in these loading operations was behind the sleigh. It consisted of one horse and a driver. A stout gin pole, 8' in height, was braced by two chains or ropes to trees or stumps. At the top was a steel pulley through which the decking line was inserted. A whiffle tree at the level of the harness had a trip hook on a handle and, at the other end of the line, was a sharp hook, one inch in width. Once the hook was made fast, the top loader would give the signal to the one horse driver to go ahead and, when the log was delivered to where he wanted it on the load, he would call to the driver "Back down." The driver would drop the trip hook and, by a twist of the arm, he would send the decking line in a rolling motion back to the sleigh. Another twist by the top loader, and the line would be delivered to the ground man. Both driver and horse would then trot back to the sleigh to pull another log from the skidway to the sleigh until the work of loading was completed.

The helper would line up the log for the ground man, who would then pass the decking line around the log and throw the hook end to the top loader, who would, in turn, insert the hook on the nearest log where the incoming one would be placed by a hard blow from the end of his

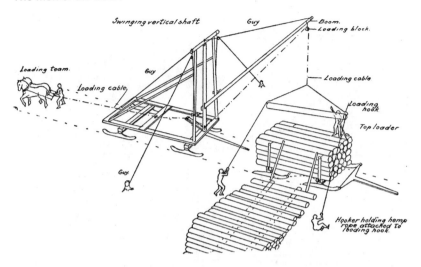

Fig. 5. An end-jammer operated by a team, typical of operations in Minnesota. From Pulpwood Hauling with Horse and Sleigh, WSI #706 (B-8) *(Woodlands Section, Cdn. Pulp & Paper Assoc., 1943), 114.*

cant hook. The hook would be disengaged by a side blow of the cant hook when the log had reached its destination. The ground man in putting the decking line around the saw log at the foot of the skids would sharply evaluate the approximate difference in weight between the butt and top end of the log. By placing the decking line at the proper distance on the log; it would roll up inside the line to the sleigh without requiring too much direction. He would, however, be ever alert and ready with his trusted, fastened cant hook at the end of the log, to advance or retard it, and thus keep it in line with the sleigh. On handling large logs, his helper would take care of the opposite end of the log. Agility and skill counted for much. Rarely would the ground cant hook men lose control of their logs. When such happenings occurred it usually led to injury to the top loader. When the log, however, simply turned about and fell between the skids with one end on the ground, and the other on the side of the road, this was called a "cannon."

Bushmen, like railroaders, miners and sailors, had a most figurative way of expressing themselves. Their colloquialisms were original. Here is an example: a top loader from Minnesota was taken to a hospital in Duluth with a broken leg. After it was set one of the sisters enquired from this bushman, whose name was "Happy Harry" Murphy, how the accident had happened. This is how their conversation was reported:

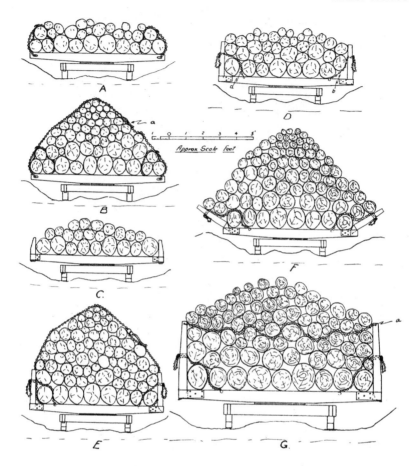

Fig. 6. Log loads of various shapes: (A) A small load on unimproved snow roads (hot logging); no binder chain; often only one log chained on each side. (B) A medium-sized load on an improved road with a binder chain (a) and beartrap binder. (C) A small load with short fixed stakes and no binder chain. (D) Also a small load with fixed trip stakes and no binder chain but on a freshly broken branch road; (a) shows log placed under stake chain and (b) shows chain hooked too short. (E) A medium-sized load on an improved snow or iced road. (F) Shows the use of inclined trip stakes to increase load. Normally a binder chain is used for such loads, but not shown here. (G) A large load between high trip stakes. A cross chain (a) between top of stakes binds the load well, but a binder chain should also be used. From Pulpwood Hauling with Horse and Sleigh, WSI #706 (B-8) *(Woodlands Section, Cdn. Pulp & Paper Assoc., 1943), 116.*

"Well, you see, I was the skyman," said Harry, "and we were shy a grounder, and there was a gazaboo come down the pike and the push took him on. The first thing he sent up was a big blue butt, and I yelled out to him to throw a Saginaw into her, but he St. Croixed her, and then he gunned her, and she came up and cracked my stem."

"I don't understand," said the sister.

"I don't either," interrupted the top loader. "I think he must have been bughouse or jiggerood."

To have witnessed such a loading operation for the first time was thrilling. It was exacting and dangerous work which was only entrusted to the most agile and experienced bushmen. One of the ablest top loaders of that period was W.J. "Sport" McGee. He had established a record at Camp 2, along the Arrow River, in December, 1900, by completing a load in 12 minutes. He was a small man, but wiry and quick as a cat on his feet. Later modern equipment was introduced in logging operations which eliminated the corner-binds and wrappers and the skill in putting up a load of logs. With mechanized innovation there disappeared from the scene one of the most fascinating groups of men in the logging industry, the famous "cant hook men" — they were gone forever.

Chapter Three

The Era of the Broadaxe

At the time that the Canadian Lakehead cities were surveyed and opened to settlement, in 1871, coniferous timber abounded every-where throughout Northwestern Ontario. These forest stands were to be of exceptional value at a later date when our railways were under construction.

Except for some fine stands of both white and norway pine south of the Canadian Pacific Railway and along the border lakes, already re-ferred to above, the forests produced tamarack, spruce, balsam, jackpine, cedar, aspen and white birch. There was a growth of sugar maple on the southern slopes of Mount McKay, some remnants of which can still be observed. The lowlands in Thunder Bay district and along the railway lines were studded in that early period with tamarack of the highest quality. It was used by the pioneer citizens of the Lakehead as fuel and taken out by contractors as railroad ties, and as piling in the early con-struction of wharves and elevators along the waterfronts of Fort William and Port Arthur.

To provide shipping facilities in 1873, the Federal Government built a short dock at Prince Arthur's Landing, now Port Arthur. It was not a very pretentious wharf. It was built in an "L" shape, and early steamers and schooners, bringing freight and passengers to the Landing, could seek its protection in heavy weather. Later, in the early 1880s, when the Canadian Pacific Railway had chosen Prince Arthur's Landing as their Great Lakes passenger and package freight terminus, this public dock was taken over by the railway company, and extended to its full length with a freight shed added. In its original form, the C.P.R. dock was the first to have been built at the Canadian Lakehead by the government.

Until the advent of railway construction on a large scale at the Ca-nadian Lakehead, private wharves were still the order of the day, as elsewhere along the waterways of the Dominion. To meet the growing demand of incoming freight to Prince Arthur's Landing and the ship-ping out of silver ore, the progressive merchants of the town constructed a number of private docks. The first one of these was built in 1875 by Thomas Marks & Company on the site where the No. 1 Canadian National Railway dock was later erected. The construction of this private wharf coincided with the building of another large dock by Oliver & Davidson for the use of the Canadian transcontinental railway (later the C.P.R.) at West Fort William. The construction of this railway to Winnipeg began on the 1st of June, 1875, and was the occasion of a ceremony when the first sod was turned over in the presence of a large assemblage of the citizens of Prince Arthur's Landing, and the new settlement along the bank of the Kaministiquia, known as the Town Plot. The dock at West Fort William was used almost exclusively for some years by the contractors along the C.P.R. to bring in their supplies, rails, and other equipment. Eventually, upon the completion of the C.P.R. from Montreal to the Pacific Coast, it was decided to consolidate the two terminals at the Canadian Lakehead. The C.P.R. purchased considerable lands on the left bank of the Kaministiquia River from private owners, and from the Hudson's Bay Company in and about their old fort. Elevators, freight sheds, shops, a round house and a coal dock were all established on their new property. West Fort William was abandoned, and Port Arthur for some time became the Canadian Pacific Railway passenger and package freight terminus.

Other docks had been built at the Port Arthur waterfront, including one for G.O.P. Clavet, constructed in 1882, followed by one for Smith & Mitchell in the same year. Other wharves had their origin in the early 1880s: the Davis, subsequently the Walsh Coal Dock; the Lake Superior Dock, eventually the Fish Dock; and the No. 5 Dock, constructed by the Thunder Bay Forwarding and Elevator Company, headed by Thomas Marks, which was the most modern of all the wharves at the time. The last enterprise was to cope with the growing need of the time for exporting silver ore during the silver boom of Thunder Bay, when Port Arthur was one of the outstanding mining towns in Canada. The P.A.D. & W. and the Canadian Pacific railways were completed at the time that this modern wharf was constructed and provided with a railway spur. In 1898 all of Port Arthur's private wharves were taken over by McKenzie & Mann for their modern Lakehead Terminus.

The Piling Industry

The construction of these early wharves necessitated a considerable amount of piling timber both in Port Arthur and Fort William. Who the first pile-driving contractors were, is hard to establish. It would seem,

however, that the Vigars brothers, Richard and William, were among
the earliest. These two enterprising young men were from Bruce Mines
and, after a stay of a few years in Marquette, Michigan, where they
engaged in some public works, they landed in Port Arthur in 1876, and
immediately organized a pile-driving contracting firm known as Vigars
Brothers. Their original equipment was fairly primitive. The motive
power used to pull up the tripping hammer was furnished by a team of
horses. Records indicate that they were the first to introduce steam
power in pile-driving, thus considerably speeding up their work. The
two brothers engaged in construction work for many years afterward,
until they erected a sawmill adjoining No. 5 Dock, near the later location of Oscar Styffe Ltd. This sawmill they operated until 1905. Subsequently, in partnership with J.A. Little, Richard Vigars organized another pile-driving firm under the name of Thunder Bay Harbour Improvements Company Ltd. He was again in his element in pile-driving
and foundation work for elevators and railway dockage facilities. Some
of the largest contracts at the Canadian Lakehead were undertaken and
successfully carried out from the time of the firm's inception in 1908,
until Vigars withdrew from the management in the early 1920s. He
lived to a ripe old age and died in Port Arthur in retirement.

Richard Vigars, "Dick" to his intimate friends, was a rugged individualist, a "hale fellow well met," and had his residence on Court Street
back of the old Post Office. It has since been converted into Sargent's
Funeral Home. If not pretentious, it was one of the fine homes of its
time in the city. Vigars was a proud man and always well attired, wear-
ing a swallow-tail coat most of the time, even when on the job. He
carried himself well. On Sunday evenings, going to Trinity Church, he
usually wore the conventional Prince Albert attire and the silk hat (the
stovepipe), and carried a cane to complete the ensemble. Richard Vigars
had many friends; he had a stock of good stories and was a good conver-
sationalist. Being a good club man, and quite familiar with the contents
of the commissary, he was rather popular with the members of the fair
sex. Public spirited, Mr. Vigars served on the old Board of Trade as
president and executive member, on the town council, and was mayor
for one term, 1905-06. He is entitled to rank as a pioneer public works
contractor at the Canadian Lakehead as well as a logger and lumber
manufacturer.

The railway companies did some pile-driving of their own in the
early days of their construction work, but other firms, who were ac-

tively engaged in pile driving, were Carpenter & Company of Fort William and Pease Brothers of Port Arthur, later taken over by Stewart McKenzie of Fort William.

Tamarack pilings were almost exclusively used during early construction. Jackpine and spruce came into use at a much later date when the tamarack stands died out through an outbreak of the larch and sawfly. The cutting and hauling of the timber for piledriving was a very substantial industry. Some of the pile timbers were cut along the shores of the various bays on some of the islands and were towed across the lake. Others came by rail, but a large number were hauled in by sleighs. It was a cheerful winter scene in the first decade of the century to see these dozens and dozens of teams of horses coming along our various primitive sleigh roads and highways with two piles on their sleighs in lengths ranging from 40 to 75 feet. This pile timber crop was of great help to the early settlers who were able to acquire some cash during otherwise unproductive winter months.

Although many of these pile timbers were purchased from individual settlers in townships immediately adjacent to the Lakehead, some of the contractors, pulpwood operators and small sawmill owners did take out a considerable amount of pile timber themselves, in some cases financing smaller contractors. Among these were Graham & Horne, Hammond & McDougall, Leonard St. Jacques, and John A. Whalen. Taylor and Mackie had extensive operations all along the P.A.D. & W. with modern loading equipment at a number of spurs to facilitate the task of loading these piles onto cars. Bruce Morrison was taking out pilings from private property at the mouth of the McKenzie River in 1915. C.W. Cox, who had been taking out pulpwood in 1911, '12 and '13 from some of his patented mining timber lots along the shore of Shesheeb Bay, obtained a large contract for timber pilings in 1915, and, during the following few years, took out a large number of piles for Barnett-McQueen. These were used in the construction of the first grain elevators to be built at the north end of Port Arthur's harbour front.

It has been conservatively estimated by competent authorities that, during the period from 1870 to 1930, no fewer than 1,000,000 pilings were used in and about the Lakehead district and along the railways in Northwestern Ontario. These piles had to conform to a certain standard of thickness at both top and bottom. The standard specification called for a 7" top and a diameter of 12," three feet from the butt, and all had to be cut from green timber. Depending on the depth to solid hardpan, which varied along the entire waterfront from West Fort William to Bare Point, an average length would seem to have been 40 feet. Some of the pilings, of course, were less than 30 feet, others ranged from 60 to 70 feet, depending on the need of the local ground. An average 40-foot length with 15" butt can be used as a fair estimate. This would mean, transferred into lumber, that total production would have equaled 200,000,000 feet, board measure.

By 1930, heavy construction of elevators and wharves was practically all completed, and the various areas which had contained such beautiful pile timber were completely denuded. The depletion of these forest stands by man had been further intensified by fungus, disease and destructive forest insects which struck with a vengeance at the tamarack trees beginning with the late 1890s and ending in 1909. Mute evidence of that insect and fungus destruction are the dead spires and some of the fallen tamarack timber among the trees which had been left uncut.

Nature, however, has a way of reproducing its forest life, and along our railways and highways healthy young stands of tamarack are to be observed. If left to grow and if protected from fire, there should be another crop to gather in fifty to seventy-five years. The preparation and transportation of piling timber was a factor in the local economy, but, important as it was, it fades into insignificance when compared to that of the railway tie operations.

The Railway Tie Industry

The construction of the railway to Winnipeg began from West Fort William in 1875, and was then a government undertaking. It was referred to as the "Lake Superior section." It was turned over to the Canadian Pacific Railway Company in 1881. The C.P.R. began to operate on it a year later from Fort William to Kenora. There is no record of who were the first tie contractors. The original railway contractors included Harry Sefton, Thomas Cochrane, I. Ward and Company, Oliver and Davidson, and Conmee & McDonald. It is possible that some of them took out their own ties.

Under the energetic leadership of William Van Horne (subsequently Sir William), then general manager of the C.P.R., the construction of that great railroad was pushed along with method and speed from the time that he arrived in Winnipeg, on New Year's Day in 1882, and established his headquarters. On his recommendation, the company immediately undertook the construction of an east line from Port Arthur to North Bay. Here it was to be linked with the Canada Central Railway, which was operating a line from Brockville to Callander, Ontario, near the present city of North Bay, and a branch line to Ottawa. The Canada Central railway was subsequently brought into the C.P.R. system, and Duncan McIntyre, its president, was made a director of the company. McIntyre Street in Port Arthur was named after him.

That great project created a heavy demand for railway ties to be used both east and west of the Lakehead. According to Harry Nicholson, a pioneer Lakehead chronicler, hemlock ties were brought in from Georgian Bay ports to Heron Bay and Rossport by steamers and schooners. But, from the early 1880s to 1888, by far the largest proportion of the ties were supplied by local contractors from Northwestern Ontario. The first tie contractors of whom there have been any records were Nagel and Egan, and McFarlane and Corbett. (Corbett Creek was named after a partner of this firm). Others who followed in sequence were James Conmee, James Whalen, John King, Philip L'Abbe, Joseph Malenfant and Oswald Hacquoil. Once the construction of the C.P.R. was completed, the demand for railway ties dropped very considerably, as did the price. Ties were being delivered then by contractors at 16¢ a piece at various C.P.R. spurs or loading ladders. The tie makers were getting 5¢ and 6¢ a piece.

In 1898 the construction of the Ontario and Rainy River Railway, promoted by Mackenzie & Mann, who had received a generous cash subsidy from the Federal Government, had begun from both ends, namely Winnipeg and Port Arthur. By 1900 its construction was in full swing, and it was completed in the fall of 1901. It became known as the Canadian Northern Railway, now the Canadian National. The construction of this road created additional demand for railway ties, pilings and bridge timbers. North America was just emerging from the depression of the 1890s, and prices on all commodities, including bridge timbers, lumber and railway ties were still low. That undertaking of MacKenzie and Mann was hardly completed when the Grand Trunk Pacific Railway obtained a charter from the Federal Government, and, under a contract by which they were to receive cash subsidies and vast areas of Crown lands, the immediate construction of a third transcontinental railway was begun. Sir Wilfrid Laurier, Prime Minister of Canada, turned the first sod in early September, 1905, at West Fort William. He had stopped off at the Lakehead for that purpose, returning from a historic western

trip where he had presided at the opening of the first legislatures of both
Alberta and Saskatchewan, which had recently been granted their provincial autonomy.

The first two decades of the present century may be referred to in
our annals as the period of the railway construction economy. It was a
time when provincial or federal governments subsidized railways, granted
charters for the construction of railways, built railways, merged railways, or took over and operated railways. During this period of railway construction, the demand for railway ties was ever present. Large
and small contracts were handed out to whomever was in a position to
make delivery of railway ties.

The following tie contractors appear in chronological order: Cap-
tain Herbert Shear, who operated along the P.A.D. & W. Railway; A.G.
Seaman, who bought ties along the P.A.D. & W. and produced tie tim-
ber along the main line of the Canadian Northern Railway to Fort Frances;
Northern Construction Company, Kashabowie; Graham & Horne at Mine
Centre; John King, then a prominent merchant of Fort William, who
had extensive tie timber operations along the C.P.R. at English River;
John A. (Jack) Whalen, who took out ties at Shabaqua; John West and
W.F. Fortune, a Port Arthur merchant, who were taking out ties near
North Lake on the P.A.D. & W.; Greer Brothers (Joe and Jim), who
were also buying ties, as were Crockett & Tharle Ltd., from the settlers
and small operators along the same railway.

Early in 1905, two very large contracts for railway ties were obtained from the Grand Trunk Pacific Railway Company by Greer Brothers of Port Arthur, and by Crockett & Tharle who had had extensive
railway experience in New Brunswick. Both opened up offices in West
Fort William. A sawmill was erected by Crockett & Tharle at Raith on
the branch line of the Grand Trunk Pacific, connecting the main line at
Superior Junction with Fort William. This mill had a capacity of 1,000
ties per day, and 10,000 feet of lumber. Greer Brothers erected two
mills along the same railway branch at Quorn and Dog River Station
with similar capacity per unit. In addition to their mill production, these
two large organizations operated a number of camps where ties were
produced by tie cutters. They also had buyers among the settlers of the
district.

It was about that time that George Mooring, originally associated
with J.J. Carrick under the name of Carrick & Mooring, Realtors &
Insurers, obtained a permit from the Crown for an extensive jackpine

stand along the upper reaches of the Arrow River. Under a contract from the Canadian Pacific Railway, he operated extensively until the limit was depleted. It was a substantial operation. His ties were floated to the mouth of the river by the Arrow River and Tributaries Boom and Slide Company, where they were boomed by the Pigeon River Slide and Boom Company, and then towed into Port Arthur by the Lake Superior Tug Company. These were all subsidiaries of the Pigeon River Lumber Company. The C.P.R. had their loading ladder for handling these ties on the north side of McVicar Creek. It is doubtful if Mooring's timber operation was ever profitable. A stubborn individual, and inexperienced in logging, he was involved in heavy litigation with the Pigeon River Lumber Company and its subsidiaries.

McDougall Brothers of Ottawa also had a large tie contract with the Grand Trunk Pacific. Ties were produced at their tie and lumber sawmills which was erected for that purpose a short distance east of Sioux Lookout at a place still called McDougall's Siding.

Judging from the records available, it can be stated that from their very origin to the end of that particular period in 1918, few operators showed any profit when their contracts for railway ties were completed. We have seen that following the completion of the Canadian Pacific Railway, the price of railway ties had dropped to as low as 16¢ delivered at railway spurs, with piece workers getting only 5¢ each. On account of the demand around 1900, the price had gone up, but only slightly. Joseph Malenfant was taking out ties for the Canadian Northern Railway at Shebandowan for 18¢ a piece. He was allowing his tie cutters 5¢ a piece in the string, or 6¢ complete. The contractor still had to do the skidding, hauling and piling at the railway spur. What was left above and beyond the total cost was his profit. Considering his overhead, there wasn't much of a margin of profit left after his obligations had been met. One of Malenfant's tiemakers, A.L. Allard, from the early autumn of 1900 to April of 1901, actually came out with a stake of $650.00 after paying his board and whatever he had required during the winter months from the camp van. He was undoubtedly one of the most skillful men of his day with the broad axe. By 1905 railway ties were commanding a better price. Overhead had gone up considerably, and prices to tie makers or piece workers were 10¢ for No. 2, and 11¢ for No. 1 ties. At this time, railway companies were paying the following prices at various spurs: No. 1 Tamarack Standard railway tie 36¢; No. 2 Tamarack Standard railway tie 34¢; No. 1 Jackpine Standard railway tie 32¢; and No. 2 Jackpine Standard railway tie 30¢.

Other contractors who engaged in producing railway ties along the three railways were A.L. Clark and C.H. (Charlie) Greer along the P.A.D. & W.; Kelly & Closs at their sawmill at Kaministiquia; James Stewart, with tie camps on both the Canadian Pacific and Canadian Northern Railways; Ray Bell and Thos. Falls, who operated along the Canadian National Railway; and Taylor and Mackie, who took out some ties. By the middle of the 1910s, Jack McKeown was a heavy tie operator along the P.A.D. & W. and eventually took over the Taylor and Mackie limits. George Farlinger, formerly a member of the partnership of Farlinger & McDonald, bridge contractors on the Grand Trunk Pacific, began to operate a tie and lumber mill at Sioux Lookout in partnership with C.H. Greer, upon the completion of his company's bridge contract with the railway. Some years later Farlinger's operations were organized under the name of the Patricia Lumber Company, of which he was the president.

Tie operation was both an interesting and a hazardous business. Many factors had to be taken into consideration. The only advantage in its favour was its quick turnover. Once a tie was produced, it was not necessary to dry it, or further process it. It was ready to be loaded onto a railway car and turned into cash. Like all other winter operations, the job was influenced to a large extent by weather conditions. Many of the large tie cuts were floated down along the drainage system of many rivers and creeks, and sometimes were held up for lack of sufficient water. In addition to all that, the railways were either in great need of railway ties or overloaded with tie timber. They might buy very heavily for two or three years, and then completely shut down on new tie material for three or four years. The operator might one year feel flush, and the next year feel like a pauper.

During the lush twenties, due to railway extension throughout the prairie provinces, ties were in great demand. Many firms and individuals produced ties for the two railway companies. Among these were: Pigeon Timber Company (not to be confused with the Pigeon River Timber Company); Don Clark; Thos. Falls; C.W. Cox; Patricia Lumber Company Ltd.; Keewatin Lumber Company Ltd.; New Ontario Contracting Company, headed by Colonel J.A. Little; and C.H. (Charlie) Greer, whose production exceeded that of the combined output of all the other contracting firms.

These tie operators were interesting individuals. Some of them were just in the business by accident without any prior training in woods operations. Most, however, were good bushmen. They were not what you would call forest technicians. They had never heard of the term "forestry engineering." They would not have known what it meant, and quite likely would have laughed at the idea of a graduate engineer from a forestry school directing their operations. The term "selective cutting" did not enter into their minds, an expression unheard of at that time. The most destructive agency of their timber limits was forest fires, with which they were familiar, and which they tried to avoid in a haphazard way. They were simply in the business to produce railway ties, and produce them they did, in enormous quantities. The tie industry was one of great importance between 1900 and 1930. Railways had not yet begun to use treated ties, and the replacement, per mile per year, required 400 new railway ties, in addition to what had been used in the original construction. There is no exact record of the number of ties which were taken out. It has been estimated, however, by competent authorities, that from 1875 to 1930, approximately 55 million railway ties were taken out of the forests of Northwestern Ontario, from Chapleau to the Manitoba boundary. The largest proportion of these ties were shipped to the prairie provinces. Reduced to board measure, this would represent a little over two billion feet of lumber. If we add the 200 million feet of lumber represented in pilings which were taken out during the same period, we arrive at an estimate of 2.2 billion feet of lumber, board measure. This estimate may be too conservative. It could have easily exceeded that amount. The forests of Northwestern Ontario have, therefore contributed most abundantly to the railway development of the Dominion of Canada, far beyond the conception that any average person may have had, whether he be a resident of the Canadian Lakehead, of Eastern Canada, or the prairies.

The tie cutters or piece workers were the skilled men among the bushworkers. They were the mechanics in woodcraft. They worked hard, but, depending on their ability to swing the broad axe, they earned good incomes. A great many of them saved their money, acquired lands and became farmers; others worked their way into business. Thomas Laprade, of Port Arthur, and the father of the famous hockey players (Edgar and Bert), was for some years a piece worker, hewing ties. He was thrifty, eventually acquired a hotel at Mine Centre, and then moved into Port Arthur, operating a hotel under his own name. Other famous tie makers were Asarius Lazare Allard, who became a tie contractor and pulpwood operator, John Moline, Rene Ysenbeart and Frank Kauzlaric, to name only a few. The production of ties is now practically a sawmill operation. Tie-making by hand has become a lost art.

The P.A.D. & W. and the Lumber Trade

Reference has been made to the Port Arthur, Duluth and Western Railway, familiarly referred to as the "Pee Dee." The pioneer merchants and contractors at the Canadian Lakehead were men of enterprise with an abounding faith in the future of Northwestern Ontario. They were ever ready to back up their faith with whatever capital they could legitimately spare, and with whatever subsidies they could obtain from the government.

The silver mining boom of the 1880s in the Thunder Bay area coincided with the discovery and development of the iron ranges of Minnesota. To assist in the transportation of supplies to the mines, and of minerals from the mines to the docks of Port Arthur, a group of enterprising citizens conceived the idea of building a railway with the additional objective of tapping the rich iron lands of Northern Minnesota. There was also the possibility of settlement in the Whitefish Valley, but this was the least of the considerations that prompted the promoters to organize this rather unique railway. Its construction began in 1884 but, for the next few years, only a few miles were built. The line was not completed to North Lake until 1889. This railway, also subsidized by the government, began to operate with one wood-burning locomotive, called "Old Black Auntie," which had been bought from the Grand Trunk Railway, possibly one of the earliest engines to have been used in Canada. She was a sight to behold, coming along the old "Pee Dee," puffing out sparks from her wood fuel ovens.

If the road had been originally ballasted, it showed no sign of it by 1900, when it was taken over by Mackenzie & Mann Ltd., subsequently the Canadian Northern Railway. They continued to operate it until 1936 when the rails were taken out. It came to be known affectionately as the "Poverty, Depression and Want Railway," for the Port Arthur, Duluth and Western Railway, was poverty-stricken throughout its whole career. There was not a single station along the whole road until 1907 when one was constructed at Silver Mountain. An unthought of objective, namely the timber along the right-of-way, became the determining factor in operating this railway from 1900 until it was dismantled.

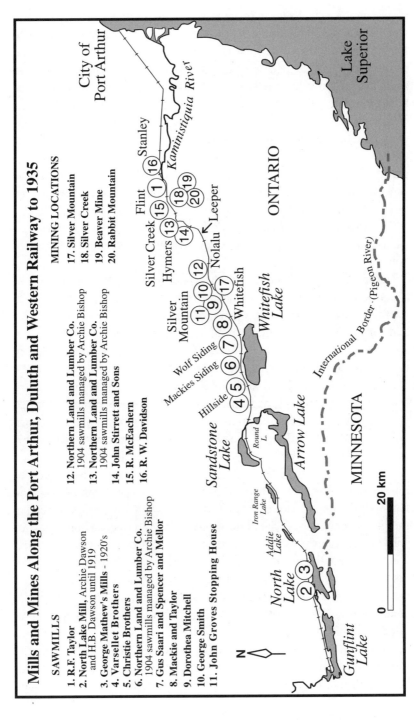

Map 4 Mills and Mines Along the Pee Dee Railway to 1935

Except for the logging trains, there were only two regular trains a week which made the run from Port Arthur to North Lake. They stopped for half an hour about noon at Hymers, in order to give the crew and passengers a chance to have a meal at the Hymers Boarding House, and again at Sandstone Lake where John Grove kept a stopping house. Here the train crew, and whatever passengers were left on the train, had their evening meal. Subsequently John Grove settled at Silver Mountain Siding where he erected a substantial one-story log cabin. He catered then to both outgoing and incoming bushmen, including those employed by the Pigeon River Lumber Company at their logging camps, along the Pigeon, the Arrow and Whitefish Rivers, and those who worked for the tie contractors. John Grove's stopping-off establishment, which served meals consisting of dried salt pork, boiled potatoes, bread and tea, became known as the Little Mariaggi, named after Port Arthur's well known hotel.

A frequent guest at the Little Mariaggi was Dr. A.H. Williamson, an associate of Dr. George Wallace Brown, who had the contract with the Pigeon River Lumber Company for servicing their bushmen. Brown had constructed a small log hospital on the upper stretches of the Arrow River, but the Pigeon River Lumber Company also operated logging camps on the Pigeon and Whitefish Rivers, and across the line on the south shore of Gunflint Lake in Minnesota. Dr. Williamson had to travel a great deal from camp to camp, and on many occasions stopped at Mr. John Grove's establishment. Dr. Williamson was a *litterateur*, a wit and a poet. With time to spare he dedicated rhymes and prose to this unique stopping-off place. He used to artistically design menus which would have been a credit to leading New York hotels. Nothing was more proper, therefore, to name this original auberge of Mr. Grove's the Little Mariaggi. There were many chuckles when this menu was read in the column of the Lakehead newspapers. On one occasion a pair of John Grove's socks, which were hanging on a wire over the stove to dry, had fallen in with the pork in the frying pan. Dr. Williamson added it to his menu as "pork a la socks." Later, a leaky kerosene lamp hanging over the cook stove had spoiled the pancakes; it thus became "griddle cakes a la kerosene" on Williamson's menu, all of which was taken in good humour by the Grove family.

John Grove was a real frontier character. He came from the timberline of southeastern Ontario, and was brought up as a farm labourer. Illiterate, but with the ready wit of his ancestors, he expressed himself with a most picturesque vocabulary. When speaking of a veteran's land grant, he always referred to it as veterinarian grant. It was indeed an amusing experience to listen to his discourse on his problems of the day.

On the Pee Dee line, it was considered a lucky run if the train only went off the track twice or three times between Port Arthur and North Lake. The crew's best act was to put the train back on the rails, at which task they were as patient and as skilled as a modern garage mechanic is in putting a new tire on the rim of an automobile. The train, of course was never on time, except when it started out in the morning at 9 o'clock. The run to North Lake used to take all day, and sometimes a good part of the night, and the return journey was just as long. The train might arrive in Port Arthur at any time from 3 o'clock in the afternoon to 7 or 8 o'clock at night. Among the pioneer railwaymen of Port Arthur who served their apprenticeship, or were connected with that celebrated railway, were George McLeod, engineer, Pete Whalen, who was the first conductor, and Adolphe Bolduc, who succeeded Pete Whalen when the latter was moved to the main line and put in charge of the first passenger train that ever traveled on the Canadian Northern from Port Arthur to Winnipeg. Other trainmen were Elias McGee, Joe Boulay, Thomas Thynne, Jim Switzer, Joe Vance, William Gilmore and Jack Lalonde.

This unique railroad rendered a great service to the district by carrying settlers and their effects, as well as by taking woodsmen to the various logging operations that this line served. But its greatest contribution was in bringing out the products of their toil and skill. It is doubtful if any other railway of similar limited mileage in Canada has ever transported so much cordwood, ties, piling, sawlogs, telephone poles, cedar posts, lumber and pulpwood in a similar span of time. The Pee Dee is entitled to some recognition for the part it has played in the development of the district.

The Pee Dee Railway at last came into its own. A number of small sawmills were erected all along the line from Stanley to North Lake. Some of the mill operators, in addition to cutting lumber, dimension or square timber, produced ties, but not in significant number, since it was still the era of the broad axe so far as tie making was concerned, particularly during the first ten years of the century. Most of the operators confined themselves to cutting lumber.

The Northern Land and Lumber Company Ltd., managed by Archie

Bishop, was formed in 1904. For a few years they operated three small sawmills along the Pee Dee Railway, one at Nolalu, one at Hymers, and one at Mackie's Siding. Captain Herbert Shear, who organized this venture, was a man of many parts. A southerner, scholarly, phlegmatic, tall, dignified, and an able businessman, he played a great part in early development along the Pee Dee. During the silver mining boom days he was captain of the Badger Mine, a title which he held to the end of his life. Subsequently he moved to Silver Mountain and operated the East End Mine. At the same time, he went into the tie business and then launched out as a lumber operator. It is unclear whether or not his mills were the first along that railroad, but they were among the earliest in operation. Archie Bishop was still manager of mill operations when the merger, already referred to, was consummated between Vigars Lumber Company and the Northern Land and Lumber Company. They decided to abandon their Pee Dee operations, which were sold to Archie Bishop, who operated them with his brother-in-law, H.B. Dawson of Port Arthur, until about 1919, when their North Lake mill was burnt out. Bishop then withdrew from the lumber business. He died in January of 1945.

Between 1904 and the mid 1920s, a number of other mill owners along this railway line and in Whitefish Valley including: R.W. Davidson, Stanley; R.F. Taylor, Flint; R. McEachern, Silver Creek; John Stirrett and Sons, Hymers; John Jacobson, Leeper; Varsellet Brothers, Hillside; Christie Brothers, Hillside; Dorothea Mitchell, Silver Mountain; George Smith, Silver Mountain; Mackie & Taylors, Whitefish; as well as Gus Saari and Spencer & Mellor, Wolf Siding. The partnership between Spencer & Mellor was dissolved in 1907. For some years afterwards Charlie Mellor was associated as secretary-treasurer with Greer Brothers, large tie contractors along the branch line of the old Grand Trunk Pacific Railway that ran from Fort William to Sioux Lookout. Subsequently he bought out the Clavet Feed Store on Court Street. In time, he became one of the opulent grain merchants at the Lakehead.

George Matthews of Fort William also operated a mill near North Lake for a few years in the early 1920s; in time it was burned out. The Thunder Bay Lumber Company took over uncut timber limits from the Pigeon River Lumber Company at Arrow Lake in 1926, and thereafter operated consistently under some arrangement with Russell McKechnie, averaging an annual cut of 2,000,000 feet a year. This can be considered as one of the major operations along the Whitefish Valley. Two of the largest operators of the early period were John Stirrett & Sons and

Parrott & Schram. Between these two organizations a very considerable amount of lumber of all kinds was cut. They shipped their lumber to Lakehead mills to be further processed. Early in the First World War these two firms were shipping a considerable amount of poplar lumber to the Chicago market. In the summer of 1915, John Stirrett & Sons shipped out a million and a quarter feet of poplar lumber and an equal amount the following winter, in addition to half a million feet being shipped by Oliver Schram to the Rathbone Company of Chicago, box manufacturers, at a price of $16.00 per 1,000 feet over the rail of vessel. With the collapse of the lumber market after the First World War, Oliver Schram of Parrott & Schram, abandoned their operations. John Stirrett & Sons moved to the Dog Lake area where they cut a considerable amount of lumber. Both these firms therefore can claim the credit of having pioneered the manufacturing and exportation of poplar lumber.

The amount of forest products taken out along the Pee Dee has not been recorded. It came almost exclusively from homesteads and old mining claims. Taking everything into consideration, fuelwood, saw logs, railway ties, piling timber and lumber, from 1900 to 1926, tens of millions of board measure can be estimated. The Whitefish Valley is now a well-settled community with good highways, well-operated farms, and attractive villages and countryside thanks in large part to the Pee Dee.

Pioneering and Privateering

The first mining boom on the north shore of Lake Superior occurred near the middle of the 1800s. A feverish period of mineral exploration was occasioned in the 1840s by rich finds of copper and iron ore along the south shore of the lake. Individuals and groups of individuals from Upper and Lower Canada were granted tracts of lands along the North Shore having a two-mile water frontage, five miles in depth, an area of ten square miles for each location. In 1845, 160 such locations were applied for from the Government of Upper Canada, all the way from Sault Ste. Marie to Pigeon River, bordering on the shores of bays, islands and straits. At that time, these locations did not carry the rights to silver and gold, which were still the property of the Crown. They were limited to copper and iron ore. Practically the whole of the North Shore was under application for mining properties.

In 1846, twenty seven of these locations were granted to and were reported upon by W.E. Logan, afterward Sir William Logan, chief director of Canadian geology. These locations actually represented a total area of approximately 270 square miles. Most of these mining properties, which eventually became patented, were located from Heron Bay to the international boundary. They were all patented under individual names. Beginning from Pigeon River, they were registered as follows: 1. James Stuart (this property is still called the Stuart Location), 2. James B. Forsyth, 3. O.D. McLean, 4. W.B. Jarvis (now Jarvis Island), 5. John Prince (the Prince Location), 6. Charles Bockus and Donald Ross, and 7. George Deberats. These formed the first group. Starting from Thunder Cape Eastward, the other locators were: 8. Joseph Woods (the Woods location of Silver Islet fame), 9. Stewart Derbishire, 10. Abner Bagg and Stanley Bagg, 11. John Ewart, 12. W.H. Merritt, 13. S.J. Lyman, 14 James Ferrier, 15. S.B. Harrison, 16. James Hamilton, 17. Peter McGill (whose father had founded McGill College, Montreal, now McGill University), 18. R.J. Turner, and 19. James Wilson. Locations 15 and 16 were on St. Ignace Island and No. 18 was on Simpson Island. These comprised all of those properties located in what is now the District of Thunder Bay.

Beginning with Michipicoten Bay towards Goulais Bay, patents were granted to: 20. Charles Jones, 21. Angus McDonell, 22. Thomas Ryan, 23. E. Arthur Rankin, 24. Edward Ryan, 25. John Douglas, 26. Alan McDonell, and 27. W.C. Meredith — all names of significance in Canadian history. These unusually large grants were the subject of considerable criticism at the time, but it was contended by the owners that prospecting for minerals along the North Shore would be expensive and hazardous because of its remoteness from eastern settlements and that, because of limited favorable weather, fairly large tracts of land would be required to justify the outlay. Most of these locations were then grouped and merged by syndicates or companies, under the names of the Ontario Land & Mining Company, the Montreal Mining Company, and the Quebec Mining Company. One of the locators, John Prince, carried out considerable exploratory work on his property at Prince Bay, and the Montreal Mining Company performed some preliminary exploration on St. Ignace Island, but without success.

Further explorations on these locations were practically forgotten from 1847 until 1868, when a rich silver find was made on a tiny island bordering upon the Woods property. This became the famous Silver

Islet silver mine. This discovery led to another mining boom. By this time the Mining Act had been amended and silver and gold were now the property of the locator. New locations were staked on the banks of the Pic and other rivers, as well as along the shores of Nipigon Bay, Black Bay and Thunder Bay, none of which proved to be of economic importance. In the early 1880s, the District of Thunder Bay was passing through another period of unusual mining activities, a mining boom which led to the discovery of Rabbit Mountain mine, Beaver mine, the two Silver Mountain mines and the Badger mine. Many more locations were staked and patented. It was from these various mining locations that early timbermen conducted their operations and exported their pulpwood to American mills.

By the early 1890s, settlers were beginning to take up homesteads in well timbered areas where arable land was also present. Late in that decade and into the next, free homesteads were granted to veterans of the Boer War. Taken as a whole, therefore, there was at the dawn of the century a considerable area of lands under patents, from which pulpwood could be obtained for export. All patented lands, prior to 1869, carried timber rights; including rights on such species as norway, white and jackpine. In 1869, however, the Timber Act was amended so that timber rights on patented lots included all timber *except* norway, white and jackpine. Jackpine was not then considered of any value; it was called the weed pine of the forest. Spruce timber, however, was then becoming valuable on account of the expected demand by American mills for pulpwood, and by large contractors at the Lakehead for piling timber. All of the above properties, therefore, acquired enhanced values.

Timber operators at the Canadian Lakehead were not devoid of imagination during the 1910s. Reasoning from faulty premises, they seem to have arrived at conclusions most suitable to their own. "Original mining locations," they must have argued, "carried timber rights with no stumpage dues and with the privilege of exporting pulpwood to American mills." Many locations were held by large companies, who insisted on collecting stumpage dues. Moreover, their mining properties were becoming denuded of commercial timber; therefore, why not stake new mining locations in fine stands of virgin spruce timber, and thus own the land and the timber.

Such must have been their deductions, and to realize how well they succeeded one only has to look at the map of Thunder Bay District from

1911 to 1920, when vast areas of our most accessible and finest spruce forests were staked out by these timber pirates, and large quantities of spruce timber taken out to be used in pilings at the Canadian Lakehead, or cut into pulpwood for export to American mills. In each case, stumpage dues were neither assessable nor collected. Moreover, so far as pulpwood was concerned, it was the equivalent of shipping it out of the country in contravention of the manufacturing clause of the Timber Act. As a result, the Crown was deprived of rightful revenue, and the manufacturing clause of the Timber Act was effectively nullified. Eventually, on March 26th, 1918, the Act was amended so that no further mining claims made after that date, carried timber rights. Sufficient fine spruce timber, however, had been secured by this subterfuge to satisfy the cupidity of these rapacious timber buccaneers. And ever since, pulpwood for exportation has been produced from these mining lots and other patented land areas.

The term "Timber Thieves," may or may not have originated during that period, but it was often heard in the streets of the Canadian Lakehead cities. Actually these loggers were not thieves or pirates in the full sense of the word. Privateers might be a better term, since their depredations were eventually legalized by the Department of Lands, Forest & Mines. Privateering became the order of the day. A pioneer captain of the Lakehead, who made a specialty of staking, reporting and patenting these timber lots for timber operators at Port Arthur and Fort William and for individual groups of speculators, had a small power and sail craft which he used for the purpose and which he properly named "The Pirate."

Another well-tested means of getting free timber from Crown or private lands was by disregarding the lines; that is, by trespassing. A timber contractor would start cutting, say on a 40-acre patented lot located at the edge or in the centre of a fine Crown forest reserve. He would cut all around the lot, completely ignoring the line, and report all his pulpwood and pilings to the local Crown timber office as being cut out from his mining location. This method of plundering public or private domains by timbermen is as old as the logging industry itself. It was done in the District of Thunder Bay. Jim Taylor, of P.A.D. & W. fame, a piling, tie and pulpwood operator, was asked by Justice Frank R. Latchford of the Royal Timber Commission sitting in Port Arthur in 1920: "If you were cutting pulpwood on your own location and saw fine spruce timber across the line, what would you do?" "I'd wheel right into her," replied the grizzled logging veteran.

From the time that Sir James Whitney formed his cabinet in February, 1905, until late in 1911, his strong man as Minister of Lands, Forests and Mines was Frank Cochrane of Sudbury, founder of the Cochrane Hardware Company. His early training had been in railway construction and timber operation. He knew timbermen and kept them in check. A capable business executive, and possessed of strong character and unimpeachable probity, he gave his department a strong, honest and progressive leadership. Unfortunately for his department, he was called to Ottawa in November, 1911, to take over the portfolio of Railways and Canals in the Conservative administration of Robert L. Borden, whose party had recently been swept into power by a large majority on the reciprocity issue.

William Hearst, the member for Sault Ste. Marie, who subsequently became Ontario's Premier, was put in charge of the Department of Lands, Forests and Mines. He was succeeded by Howard Ferguson a "hale fellow well met" who, in addition to administering his own department, a full-time job for any minister, was also obliged to give a great deal of time to the Department of Education, whose minister was laid up with a serious ailment which eventually carried him to his grave. To make things worse, the veteran Premier of the province, Sir James Whitney, was permanently incapacitated by a stroke early in 1912. He died in the autumn of 1914.

William Hearst was chosen to succeed Whitney, and from then until the general election in the fall of 1919, the province was directed by what may be called a "Pink Tea Administration" just when it was in need of strong leadership. Laxity and administrative abuse kept creeping into the Department of Lands and Forests. There had been no lack of prompting and criticism from within and without. In November of 1914, W.T. McEachern, president of James Hourigan & Company Ltd., who were operating under a special timber permit on the Black Bay Peninsula, and A.J. McComber, their solicitor, called on the Minister of Lands and Forests in Toronto and warned him of the trouble that would inevitably follow if some drastic action were not taken forthwith to put an end to the growing abuse of forest regulations. When they told him that vast areas of the finest forest lands were being staked under the pretext that minerals had been found and that the assessment work required by the Mining Act had been performed, for which they had to take an affidavit in the presence of the mining recorder (then J.W.

Morgan), Howard Ferguson merely laughed-and turning to Alex McComber, said: "Isn't Port Arthur the place where they fabricate affidavits?"

This vicious system made it difficult, if not impossible, for legitimate, honest and responsible timbermen to operate. They had to pay stumpage dues on sawlogs, piling timber and pulpwood; the forest buccaneers got theirs free of Crown dues. The Crown timber agent, J.A. Oliver, had warned the deputy minister; private citizens had written letters of warning to the minister himself, and even to Donald M. Hogarth, M.P.P., in service overseas. These warnings did not seem to have much effect until early in 1918 when one of the "timber pirates" happened to write to a friend, telling him how easy it was to obtain these timber lots. He had also written another private letter, on a different subject, to the Honourable Frank Cochrane, then at Ottawa. Unfortunately, he had put the wrong letter in the envelope destined for Mr. Cochrane and naturally when the latter learned what was going on, he immediately took steps with the Ontario government to have the abuse remedied. On March 26, 1918, the Mining Act was amended to reserve all timber on mining property for the Crown.

This much-needed reform had come too late. The ship of state was heavily laden with barnacles of all sorts, and when the storm broke in the fall of 1919, it finally foundered. The administration was swept out of power. During the campaign, a great deal had been heard across the province about the spoliation of our public domains. Promises had been made by Liberal speakers that, if elected, they would appoint a royal commission to inquire into the administration of lands and forests, and would introduce legislation to put an end to defrauding of the province of Ontario of its stumpage revenue. A Farmer-Labour (the United Farmers of Ontario or UFOs) coalition was entrusted with the task of administering the affairs of the province, with E.C. Drury as Premier. The Liberal leader, Hartley Dewart, K.C., at the first session of the House forced the issue on the matter of timber investigation. A commission was appointed, consisting of Judge William R. Riddell and Justice Frank R. Latchford, both of the Court of Appeal to investigate timber cutting practices in the province, particularly those involving the "Timber Ring" at the Lakehead. By the early summer of 1920 they began to hold sittings in the eastern part of the province; by late summer, these sittings had been held all the way from the boundary of the province of Quebec to that of Manitoba.

The rich province of Ontario at that time was not getting as much revenue from its vast forest lands as did the small province of New Brunswick. Therefore, the people of Ontario had a right to be inquisitive and to demand reform. The commission was accused in some quarters of being politically biased, and was bitterly assailed by local newspapers. Nevertheless, it conducted its investigation with resolution and skill. It was soon discovered that pulpwood operators were not the only violators of forest regulations. There had been gross laxity in the administration of the department. Small and large logging and lumber companies had indulged fairly freely in the sport of trespassing on Crown domains and making hip-pocket returns to the department. Many of the operators were forced to wear the sack cloth of penitence. Fines were imposed, one of them exceeding the half million dollar mark. Most of the penalties, however, ranged from $5,000 to $50,000. Like the "robber barons" of the Middle Ages, a few of these operators — but only a few of them — hurriedly and full of repentance, brought their cheques to the commissioners before they were ordered to do so in restitution for any wrongdoing. One of the timber men, who produced a large quantity of ties, an individual of a rather pious nature, naively stated to the two presiding judges that he always believed that the forests belonged to God, and therefore felt free to do anything he wanted with the trees. Since trespassing was such a common infraction among the timber operators of the period, he could have added "forgive me for my trespasses." It is not likely, however, that he would have been so willing to forgive those who were trespassing on any of the timber stands which he claimed as his own timber limits.

Actually the average citizen in Northwestern Ontario was not greatly disturbed by the revelations of fraud brought out by the Timber Commission. General Donald Hogarth, M.P.P. may have correctly interpreted the minds of his electors when he stated that "after all there was plenty of timber, and the stumpage only commanded forty to sixty cents per cord." Another leading citizen of Port Arthur, and by the way, an idealist, expressed the opinion that our trouble was that we had an overabundance of forest resources, which lent themselves to irresistible temptation and covetousness. It was, in short, a repetition of what had occurred previously in the border states at Wisconsin and Minnesota. Public opinion didn't wake up or become aroused until the timber stands of those two great states had been practically depleted.

Nevertheless, those of the operators who had submitted briefs to the commissioners, and made restitution were publicly congratulated for their good deeds, and their briefs were accepted and did assist the commissioners in recommending needed legislation. It is well to keep this background in mind when reviewing the history of pulpwood operations from their earliest origin until the year of the deluge — 1920. By this time, pulpwood for export was commanding a price of $20.00 per cord for four-foot wood. Those of the operators who had survived the inquisition were soon to be on Easy Street. Some, however, folded up; others had already collapsed. A number carried on and remained active, but a new timber aristocracy was in formation. The pirates and privateers were gone, but the wolves were coming in from over the hills.

If the Royal Timber Commission had not accomplished as much as was expected, they did succeed in arousing the Department of Lands and Forests to a greater sense of responsibility, and in obtaining increased revenue from stumpage dues. This has since benefited all the shareholders of the great province of Ontario.

Chapter Four

Pioneer Pulpwood Operators

T he early 1890s witnessed the first pulpwood operation on the north shore of Lake Superior in what is now Northwestern Ontario. The general contractors were the John Nesbitt and Company of Sarnia and the pulpwood was exported to Port Huron, Michigan. This wood was cut from patented properties along the mouth of the Pic River and was made up of spruce timber of sawlog proportions. Four years later, James Whalen, who had been doing some sub-contracting work for this firm, bought their equipment and took over the general contract with the Port Huron Sulphite & Paper Company of Port Huron, Michigan, a pioneer firm in the sulphite and pulp industry in the United States. A partnership was organized with Whalen, Richard Armstrong Hazlewood, a former Canadian Pacific Railway engineer, and James Conmee, a leading railway contractor of the time. The firm became known as Hazlewood and Whalen and they operated consistently under that name until the summer of 1898. They obtained their pulpwood mainly from patented properties along the North Shore from Pic River to and including Black Bay. Prices were low, from $3.60 to $4 per cord and profits were proportionately small. James Conmee, then a member of the Ontario legislature, withdrew from the partnership and his shares were taken over by his son-in-law, James Whalen. Conmee re-entered into railway work with another partner and for some years carried out railway construction under the name of Conmee & Middleton.

During the winter season of 1894 and 1895 another pulpwood contracting organization entered the picture. A partnership was formed consisting of J.T. Emmerson of Wells and Emmerson, Hardware Merchants, Joe Brimstone, James Meek, all of Port Arthur, and Robert B. Whiteside of Duluth. The latter, as previously mentioned, had some logging experience on the Pine River a few years previously in a sub-contract with James Conmee Company. It is quite likely that this venture was financed by Wells and Emmerson. Their timber property was along the bank of the Willow River, a small stream discharging its water into Lake Superior some fourteen miles southeast of the Pic River. Again this cut was taken from timber of sawlog dimension considered, along with that of the Pic River, the finest spruce along the North Shore.

Their pulpwood was watered and boomed in unprotected water. A heavy gale broke the boom and the results of their toil and investment were scattered along a good portion of the North Shore. With the help of Hazlewood and Whalen men, they were able to salvage a portion of their cut, but their operation ended in a heavy loss. They sold out their equipment to Hazlewood and Whalen and pulled out of timber operations altogether.

The Lakehead's pioneer merchants were enterprising individuals and seemed to have always been prepared to gamble. Despite low prices and uncertain financial returns in pulpwood operations, Thomas Marks and Company, the leading merchants of Port Arthur, were taking out a cut along Nipigon Bay during the season of 1895 and 1896 with Ed Marks as superintendent.

Up to this time the towing and general tugging work on these contracts was largely performed by Servais brothers, Captain Harry, subsequently owner of the Ottawa House, and Joe, engineer of the tug. They eventually sold their tug, the "Siskiwit," to Hazlewood and Whalen. This was the beginning of a large fleet of tugs subsequently operated by the James Whalen interests.

During the early autumn of 1898 a new firm was formed by Whalen and Hazlewood. For the first time American capital was being invested in a pulpwood contracting firm along the North Shore. This company was organized by Letters Patent dated September 8, 1898, under the name of the North Shore Timber Company of Port Arthur Ltd., with the following officers and directors: Paul Weidner, president; Alfred Kossuth Keifer, secretary-treasurer; Edward William Voigt and Otto Louis Edgar Weber, all of Detroit, Michigan. These men held the controlling interest in the Port Huron Sulphite and Paper Company. Port Arthur directors were Hazlewood and Whalen, the latter as manager at the Lakehead. For the next four years the new organization was active in cutting and exporting pulpwood. The prices were now a little higher, $5.50 to $6 per cord for four-foot wood delivered over the rail of the vessel; but still it wasn't considered a highly profitable operation. Operating costs had to be kept down to an extent that no sub-contractor did any better than break even. For example, James "Winchboat" Whalen, of Pembroke (not related to James Whalen of Port Arthur) and Charles McCarthy of Port Arthur were in partnership in taking out 8,000 cords of four foot pulpwood for the North Shore Timber Company of Port Arthur Ltd.

from patented locations at the mouth of the Black Sturgeon River at $2.75 per cord. For 12- to 16-foot logs piled on the bank of the river they were getting $2.40 per cord. During the same winter of 1900 and 1901, Ed Closs of Nipigon also had a contract with the Whalen interests to take out 14,000 cords of four-foot wood at $2.75 per cord. Count de la Ronde of Nipigon also had a contract to take out four-foot wood, but for some reason he was to receive only $2.50 per cord piled on the river bank. The North Shore Timber Company Ltd. had to finance their sub-contractors, water the wood, boom it, and deliver it over the rail of vessels for the margin that was left between their price to contractors and what they received from the mills. This particular arrangement was undoubtedly a good proposition for the mill owners at Port Huron, but did not seem profitable to the North Shore Timber Company of Port Arthur nor to their sub-contractors. The company folded up and passed into history.

The rapid development in our western provinces necessitated modern harbour facilities at the Lakehead, which meant a considerable amount of dredging in both cities. James Whalen saw his opportunity and, with James Conmee, bought an old dredge and began dredging under the name of Whalen and Conmee. Two years later the Great Lakes Dredging Company Ltd. was founded with James Whalen as president. By October 1904, James Conmee had been elected as a member of parliament and was therefore precluded from having any shares in a company engaged in federal public works. He disposed of his interests in the company to Bowman Brothers of Southampton, Ontario.

For the first 25 years of the present century James Whalen was unquestionably the leading figure in the industrial expansion and public works development at the Lakehead. In addition to the Great Lakes Dredging Company and the Canadian Towing and Wrecking Company, he organized a number of subsidiary firms. He promoted the Port Arthur Shipbuilding Company Ltd. and erected the Whalen Building (now owned by Thunder Bay Hydro); he had already constructed the Lyceum theatre in 1909. His last venture in building construction at the Lakehead was the store and office building later owned by Tomlinson Brothers Ltd.

To all these varied enterprises he gave a great deal of his time and the benefit of his unusual organizing ability. He was a man of restless energy and vivid imagination; he was a born optimist, and generous to the extreme. When he sold out his interests in the Great Lakes Dredging Company and subsidiary firms he invested heavily in a pulp and paper venture in British Columbia known as the Whalen Pulp and Paper Company Ltd. This firm had been promoted by his brothers, William and George, the latter having had some training in pulpwood operations at the Lakehead with Whalen and Hazelwood. The brothers, with the backing of James Whalen, had undertaken to pioneer a new field in British Columbia, twenty five years too soon. The undertaking proved particularly disastrous to James Whalen who never fully recovered from his financial losses. He died in 1929, in his 61st year, in his beloved Port Arthur. Of him it can be truly said: "For his many achievements, look around you." He was the real pioneer pulpwood operator in the district.

In the late 1890s and until 1903, E.A. (Ed) Carpenter, a pioneer citizen of Fort William, was taking out pulpwood from mining properties near Cloud Bay and Prince Bay. His booming ground was in Cloud Bay. John A. Oliver of Fort William was superintendent of his woods operation. In 1905 Oliver succeeded Hugh Munro as crown timber agent at the Lakehead, and took up his residence in Port Arthur. Public spirited, Oliver sat on Port Arthur's city council for many years and was mayor of the city for two terms. Eventually he was elected president of the Board of Trade (now the Chamber of Commerce). A kind-hearted fellow who was unable to say "no" often enough to his many timber friends, Oliver was heavily censured by the Timber Commission when they sat in Port Arthur in the summer of 1920. He tendered his resignation and was succeeded by his assistant, J.M. Milway.

When James Whalen abandoned his pulp cutting operations, several important Detroit industrialists and financiers were deprived of their source of good spruce pulpwood from the north shore of Lake Superior. They responded early in 1902 by organizing the Lake Superior Timber Company under the laws of Michigan, to supply pulpwood to American mills. S.T. Miller of Detroit was appointed its president. They acquired all the cutting right from the extensive mining locations of the Montreal Mining Company and the Ontario Land & Mining Company on St. Ignace Island, on Simpson Island, along the Nipigon Straits and on other parcels of private property and veteran homesteads. They opened up an office in Port Arthur in the old Ross Block, and began operations early that summer. The tug "Superior," a fairly large craft with a quarter deck, was brought up from the east and W.L. Bishop was appointed manager of the enterprise. A six-footer and an exceedingly well-built man, he was the Beau Brummell of his day. He was not a specialized logger, and it is not likely that he made much impression on the bushmen;

but he was highly decorative when walking along Water Street, particularly to and from the old Northern Hotel (later the Mariaggi). By 1905, the company was in financial difficulties and went into liquidation.

In the spring of 1906, a new organization was formed by a Detroit group to take over the assets of the defunct Lake Superior Timber Company. It was headed by Herbert H. Edward, president of the Detroit Sulphite Company and was named the Northern Islands Pulpwood Company Ltd. Arthur J. Richardson, vice-president of the Detroit firm, became general manager. They began operations during that summer. An office was opened in the Mooring Block in Port Arthur with Barney Bell, a most colourful character, in charge of the office and all purchases. They started out in a big way, buying equipment, building new camps, opening new roads and adding much needed improvements along the rivers. A large number of men were sent out to cut pulpwood and heavy operations were undertaken during that winter with big Jack "Sandbar" Murray in charge.

The following year, 1907, North America was again experiencing one of its periodic financial panics which brought low many of the substantial financial, mercantile and industrial corporations in the United States. The first to go down was the famous Knickerbocker Trust Company in New York City. By October of that year, the Northern Islands Pulpwood Company was in serious difficulties and was unable to meet its obligations. Lack of credit impaired its future operations. In response, the management in Detroit sent Walter H. Russell, a talented young Detroit lawyer newly graduated in corporation law from the University of Michigan, to inquire into the situation at the Lakehead and make a report. Russell had had some experience the year before in Canadian legal matters when he had been sent to look over some delicate land deals by the same company along the banks of the lower St. Lawrence. After making his report, he was appointed manager of the firm, and the following year took up residence at the Lakehead. He was identified with this firm until about 1911.

During this period, Northern Islands conducted extensive woods operations on the mining properties they had acquired from the defunct Lake Superior Timber Company and also from veterans' lands along Trout Creek near Nipigon and on the banks of the Black Sturgeon River. The pulpwood they obtained from these locations was cut into 12- and 16-foot lengths. These were hauled to the river, floated down, boomed at a slasher mill near the C.P.R. track, hauled up over a jack ladder, cut into 4-foot lengths, dumped into the river again to be floated down to its mouth, then boomed and towed to the boats to be loaded. Harry Dulmage, who had been associated with the company in an official capacity, became for a time a sub-contractor. Prices, however, were still low; and if the parent company made some money by processing the wood into pulp or paper, it is doubtful if the timber operation itself — cutting, floating and loading of the wood — ever paid high dividends.

The Pigeon River Lumber Company, already referred to as manufacturers of white pine lumber, were, in addition to their logging and sawmill operation, buying a fairly large amount of pulpwood from timber contractors whom they were financing, and from the settlers who had been taking up lands adjacent to the Lakehead. Two of these contractors in the period extending from 1907 to 1912 were J. St. Jacques of Port Arthur, a pioneer timberman, and C.E. Smith, formerly an engineer on the staff of the Kam Power Company. The latter eventually formed a company of his own. Meanwhile, he was operating largely from old mining lots along Prince Bay and Cloud Bay. St. Jacques took out considerable pulpwood from the north shore of Sawyer's Bay, his last operation being in 1912-13. The Pigeon River Lumber Company in turn exported this pulpwood to Michigan and Wisconsin mills. Some of it was shipped by boat and was loaded at their booming ground near their large sawmill in Port Arthur, but by far the largest amount was towed to the railhead at Ashland, Wisconsin, to be loaded onto cars. If not the largest exporters of pulpwood, Pigeon River Lumber were a factor in the buying and selling of pulp timber, and their yearly exports were fairly consistent; but it is doubtful if they ever made any real profit out of these pulpwood transactions.

By the spring of 1911, the young Michigan corporation lawyer, Walter H. Russell, was a full fledged timber operator. With the backing of Wisconsin mill owners, he organized the Russell Timber Company Ltd. with headquarters in Port Arthur and timber operations located along Black Sturgeon River. Wood was now commanding $7.50 per cord. He began his operation during that winter and by the following spring was exporting pulpwood from his newly-acquired patented timber lots. At this time, North America was again enjoying a period of trade expansion and rapid industrial development. High prices continued to remain in force for 4-foot pulpwood until 1913. By the fall of 1914, some of the American mills were making contracts at the Lakehead for 4-foot pulpwood at a price ranging from $5.75 to $6.00 per cord over the rail of vessel. The Russell Timber Company, however, continued as heavy operators and exporters.

Another company was formed early in 1914, largely inspired by Walter H. Russell, by now the most considerable pulpwood operator and exporter at the Canadian Lakehead. This was the Newaygo Timber Company Ltd., organized under the laws of Ontario, with headquarters in Port Arthur. Newaygo had received a contract from the Pulpwood Company Incorporated of Appleton, Wisconsin, which was supplying wood to the mills in the Fox River Valley of Wisconsin. From its inception until 1921, Walter Russell acted as the new company's Lakehead agent. Newaygo had acquired all the timber rights formerly held by the Northern Island Pulpwood Company, now in liquidation, from G.T. Clarkson of Toronto, trustee for the liquidators, though all Northern Island's equipment was to be taken over by the Russell Timber Company. From these old mining locations, now held by Newaygo, a considerable amount of pulpwood was taken and exported to Wisconsin paper mills

The joint operations of these two timber companies, under Russell's management, assumed vast proportions between 1914 and 1920, with an average of 75,000 cords of 4-foot pulpwood being exported annually to American mills. Some were shipped by steamer, but the largest proportion was boomed and towed across to Ashland to be loaded onto railcars. They were operating tugs of their own and for the purpose of towing their large rafts across the lake, they had leased the fastest, largest and most powerful tug seen on the Great Lakes for many years, "The Traveller." This tug was a familiar sight along the North Shore and in Port Arthur from 1914 until 1920, when the operations of the Russell interests were investigated by a Royal Commission. Officially censured, and heavily fined for trespassing and other violations of the act, the Russell Timber Company from then on operated only intermittently until the late 1920s. They never did regain their position as the largest timber exporters of the district. From the 1930s on, during the depression years, the Russell firm had practically retired as active operators; but they still controlled considerable areas of patented timber lands from which operations were conducted by other contractors, ending unfortunately in litigations. In 1943, Walter Russell completely abandoned his timber interests and was appointed police magistrate for Port Arthur, which jurisdiction extended to the eastern boundary of the district of Thunder Bay.

Other pulpwood operators included C.W. Cox who was taking out a large number of piling timber from his mining patented timber lots along the shore of Shesheeb Bay on the Black Bay Peninsula. He had been taking pulpwood from these timber lots in 1912 and 1913 for the Detroit Sulphite Company. This early venture did not seem to prove too profitable and he withdrew from pulpwood operations until 1917, when pulpwood was commanding a better price.

In 1914 the Western Contracting Company Ltd. was organized at Nipigon, with C.L. Bliss as president and A.V. Chapman as secretary and treasurer. They were buying pulpwood from settlers in the Nipigon district, and operating under a yearly contract with the Port Huron Sulphite Company of Port Huron, Michigan. They had bought and shipped a yearly average, during their operations, of 12,000 to 15,000 cords. Their first contract price for 1914 and 1915 was $5.75 per cord for 4-foot wood over the rail of the vessel. They operated until 1932, when the firm was dissolved, C.L. Bliss retired altogether, and A.V. Chapman bought out the Charles Mellor grain business in Port Arthur.

During the same period, James Stewart & Company became fairly large operators and exporters, and so did Thos. Falls, but their particular interests were railway ties. If they were not major pulpwood contractors during this early period, they did nevertheless cut pulpwood from their lands and bought some from settlers which they shipped to American mills.

By the summer of 1913, the famous Canadian real estate boom had suddenly collapsed. Banking and financial institutions had become frightened. Many large apartment and office buildings in the leading towns of Western Canada were left unfinished. By early 1914 it was evident that the country was heading for hard times, unemployment and distress. The situation was aggravated by the outbreak of the First World War in late July of that year.

Federal and provincial authorities became anxious about the unemployment situation and both were prepared to make concessions to anyone providing work by either giving special cutting rights from Crown lands with the privilege of exporting pulpwood or other timber for letting out contracts for public works. Such a permit to cut pulpwood from timber reserves was given for a two-year period to the James Hourigan Company Ltd. of Port Arthur, of which W.T. McEachern was president and James Hourigan manager. The contract with the government was renewable in two years' time, at regular assessed stumpage dues. The exportation clause was only for that period, to be reissued, if found necessary, by an order-in-council. The timber limits allotted to them were from the Black Bay reserve, along the shore of the Shesheeb Bay. A logging operation was set up, equipment purchased, a tug leased, and the famous Jack "Sandbar" Murray put in charge as superintendent. It was the last Lakehead enterprise with which he was associated; he left two years later, to settle in British Columbia among the big fir trees. Murray was never at home as a pulpwood superintendent.

The James Hourigan Company had a contract with the Detroit Sulphite Company. After the first winter's operation, it was discovered that expenses had to be heavily curtailed. They had not made any profit, and Jim Hourigan took over the woods management himself to effect some saving. They were unable, after the end of the third year, to renew their contract to export their pulpwood from this Crown reserve. Lumber being in good demand in the United States, they took out a rather heavy cut of sawlogs during the following winter. These were boomed and towed to Port Arthur and made into lumber at the old Vigars Bros. mill, which had been put in repair for this project. The entire mill production was exported to American lumber yards. It is doubtful, however, if this operation was profitable. By 1918 pulpwood was beginning to be in short supply, and was commanding high prices. In 1920 it actually reached the unheard-of price of $20 per cord for 4-foot wood. Another attempt was made by the James Hourigan Company to take a rather heavy cut of pulpwood from the Crown limit which they held on permit basis, despite the fact that their special permit to export from this limit had not been renewed. Their operation ended disastrously. They were accused of trespassing, exporting pulpwood from Crown lands, and serious irregularities since they had claimed that their pulpwood had been obtained from beach combing. As a result, the Hourigan Company was publicly censured and heavily fined by the Royal Timber Commission in June 1920. They decided to abandon their operation. Their equipment was taken over by C.W. Cox who continued to take up pulpwood for export from his patented mining lots until they were all cut out. When the Royal Commission was sitting in Port Arthur, there were good jokes going around among the timbermen at the Lakehead regarding the Russell & James Hourigan companies both of which had been so closely scrutinized and heavily penalized by the two judges. It was held that the Russell Timber Company had been mining for their timber and that the James Hourigan Company had been fishing for theirs.

Walter Russell could properly have been called the "Jean Lafitte" of this period of timber buccaneering. The James Hourigan Company completely disappeared from the scene.

By the fall of 1912, C.E. "Haywire" Smith, who had been taking out pulpwood as an independent contractor from mining locations along Prince and Cloud Bays, founded the Central Contracting Company Ltd., with headquarters in Fort William. For the next five years this organization conducted heavy and extensive operations, exclusively from new patented mining lots, mostly along the Black Bay Peninsula. In volume of business, they ranked second only to the Russell Timber Company, their pulpwood being taken out under a contract to the Hammermill Paper Company of Erie, Pennsylvania. What ability Smith may have possessed as a hydro engineer is not known, but he certainly was no logger or timber operator. Despite the fact that all this wood was free from stumpage dues — in other words "free wood" — his company went into liquidation early in 1920. Smith's operation, therefore, did not come under the scrutiny of the Royal Timber Commission.

None of the pulpwood contractors or operating companies grew rich from their toil, worries and investments in the decade before the days of reckoning in 1920. Neither had the Department of Lands, Forests and Mines, for reasons already recorded, received any revenue commensurate with the value of pulpwood exported from Northwestern Ontario to American mill yards. The American producers only had profited from the denudation of the fine spruce forest bordering the north shore of Lake Superior. Only one remedy could possibly effect a cure and that was the establishment of pulp and paper mills in the district.

The Growth of Pulp and Paper Mills in the Northwest

As early as 1908, the Honourable Frank Cochrane, then Minister of Forests, Lands and Mines, had made a vigorous attempt to bring American mills close to the source of supply. In return for building a paper mill, his department was prepared to grant substantial timber limits along any of the rivers flowing into Lake Superior from the Nipigon to the Pic on the basis of 40 cents per cord stumpage dues for spruce. His department was also prepared to give absolute water-power development rights on any of these rivers. William Scott, then secretary-treasurer and manager of the Pigeon River Lumber Company, was approached. He be-

came interested. He had some timber cruisers look over some of the
forest lands and hired a hydraulic engineer to estimate the flow of the
Pic River. Scott had had some experience in paper mill organization.
He was one of the original stockholders of the Northern Paper Company of Grand Rapids, Wisconsin (now the Consolidated Water Power
& Paper Company), and had been secretary-treasurer of the company
in the late 1890s, when paper making was not too profitable and before
the Pigeon River Lumber Company was organized. In response to
Cochrane's offer, he made several trips to the United States, consulted
with the partners of his own company, and interviewed responsible financiers in Milwaukee and Chicago. Although some interest was shown,
nothing came of his efforts. In 1912, J.J. Carrick, M.P. for Port Arthur,
with Sir William Mackenzie and Sir Donald Mann of Canadian Northern Railway fame, tried to get William Scott and his associates interested in a paper mill project at Nipigon. This was a pretentious scheme
for that time, and would have involved an expenditure of some
$5,000,000. It was felt by J.J. Carrick and his two associates that the
Pigeon River Lumber Company was the right group of manufacturers
to carry out the project. They had the experience and the equipment and
were well organized. Moreover, most of the partners of the firm at that
time were men of substance. William Scott once more became highly
interested, and made a special trip to consult with his partners in Wisconsin Rapids. However, they were not interested and the project did
not even reach the stage of provisional organization.

Late in 1912, Walter Russell and his associates submitted a proposition to the City of Port Arthur by which they would undertake a 50-
ton pulp mill on the right bank of Current River, near the lakeshore.
They required eight acres to establish a wood yard and build their mill.
At that time the property was owned by the Parks Board, which opposed the deal. A tentative agreement was submitted early in January of
the following year, to be ratified by the ratepayers, but opposition from
the Parks Board defeated it. The same property was conceded nine
years later to another small pulp mill, which will be referred to in the
next chapter.

No further attempts were made to induce paper mill firms to estab-
lish their mills at the Canadian Lakehead until the late summer of 1915.
During that year, John A. Oliver, a former mayor of Port Arthur and
Crown timber agent, was president of the Board of Trade, and A.G.
McCormick was its secretary-treasurer. At the first fall meeting of the

executive members, a plan was outlined for their approval, to have paper mill interests from Eastern Canada or from the United States establish a plant or plants at the Lakehead. The president and secretary were then authorized to send out a series of well-prepared letters outlining the forest resources of the district and the advantage of locating plants here, not forgetting the cheap and abundant hydro power available.

Scores of such appealing letters were sent to American and Canadian pulp and paper mill executives and to financial groups who might be interested in the promotion of such mills. One reply was received from A.G. McIntyre, a consulting engineer in Toronto, who represented a group of paper mills. Negotiations were opened and shortly afterward McIntyre arrived in Port Arthur to discuss an agreement on behalf of his principals, the Provincial Paper Mills Ltd. The identity of his principals, however, was withheld until the agreements with the city had been finally endorsed by the ratepayers.

When the agreement was submitted to the ratepayers, it led to considerable public discussion. The city's offer of lands at the north end of the harbour, then known as Bare Point, and tax concessions (except for public schools), was bitterly assailed from the platform by Col. J.A. Little and his small minority group. Their bark was much more dangerous than their bite, however, although for a time the advocates and promoters of the project had goose flesh. But the committee in charge of the agreement, composed of aldermen and professional and business men, as well as members of the Board of Trade, were well organized, and when the result of the voting was known, the agreement had carried by a large majority of the ratepayers. The promoters then took out a charter. The firm, known as the Port Arthur Pulp & Paper Company, began immediately to prepare plans for a 50-ton sulphite mill. This was the origin of the Provincial Paper Company Ltd., the first wood pulp mill erected and operated in the district of Thunder Bay.

I.H. Weldon, president of the Provincial Paper Company Ltd., was made president of the subsidiary company at the Lakehead, with S.F. Duncan as secretary-treasurer. A.G. Pounsford was appointed general manager, and T.R.H. Murphy became construction engineer. Building began in March of 1917, and the mill was in operation in February of the following year, with A.J. Hansky as superintendent, C.E. Gardiner as superintendent of woods operations, and for a time, Angus G. McCormick acted as pulpwood buyer and supervisor of the wood yard.

The dreams of the pioneer residents of the Lakehead had at last been realized. Since the early 1900s they had hoped for, talked about, and prayed for a paper mill in their cities. With the exception of Walter Russell's proposal in 1912, they had been fully prepared to support in a plebiscite, any reasonable proposition which would bring a mill. For some time after they had endorsed the agreement with the promoters of the Port Arthur Pulp & Paper Company, it looked as if their victory was going to be of short duration; simply another dream. The prime factor which persuaded the executive heads of the Provincial Paper Company to establish a plant at the Lakehead was accessibility to adequate timber limits to feed their mill. Two large areas were opened up by the Department of Lands and Forests for that purpose, and bids were invited for two fine tracts of spruce along the Black Sturgeon and Pic Rivers. These areas were cruised by competent timber men, and, in turn, they submitted their tenders to the department. To the great surprise of everybody, including officials of the Provincial Paper Company, their price was outbid by J.J. Carrick who had put in two tenders, one in his own name and one in the name of S.A. Marks, a timber operator from Sault Ste. Marie, Ontario. Marks had taken out pulpwood in Port Arthur which he bought from the settlers in January, February and March of 1915, but he had not before dealt in pulpwood at the Lakehead. J.J. Carrick, who had resided at the Lakehead from 1904 until the outbreak of the First World War, had attained local fame as a real estate promoter. He was in turn an alderman and mayor of the city, then a member of the Legislature of Ontario from 1908 until 1911, when he received the nomination for the federal house and was elected by acclamation. He sat as representative for the constituency in the House of Commons from the time of his election until 1917, when he failed to have his nomination endorsed by the proponents of the Union Government at the Lakehead.

By a stroke of the pen, some imagination and quite likely a bit of political skullduggery, J.J. Carrick became the concessionaire of the two finest timber limits in the Thunder Bay district. A strange anomaly it was, with the resultant spectacle of a timber speculator with no previous experience in timber operations holding vast tracts of spruce timber, with no pulp and paper mill to supply, and a paper mill company erecting a plant at the Lakehead, with no timber limits of their own.

To their credit, officials of the Provincial Paper honoured their agreement with the city of Port Arthur and were eventually rewarded. The Department of Lands and Forests subsequently opened up another reserve in the Sibley Peninsula, just across Thunder Bay. This was put on the market as required by the statute, and the tender submitted by the Port Arthur Pulp & Paper Company was accepted. This gave them a beginning. At a later date, reallocations of timber limits were made, and by the 1950s, Provincial Papers was operating from 2,500 square miles of the highest spruce timber in most accessible areas. They installed their first paper machine in the mill in 1922, the second in 1926. At the time of writing they produce 170 tons of book paper per day, as well as 75 tons of sulphite and 60 tons of ground wood. In these operations they actually use 60,000 cords of spruce, 4,000 cords of poplar and 2,000 cords of birch every year. They employ up to 750 men in their woods operation and 500 men in the mill. The first process of its kind in Canada — coating paper on both sides — is being introduced in their mill. This mill operated as a closed shop since 1919, under a contract with the American Federation of Labor and have had a successful operation since the beginning, free of labour trouble.

J.J. Carrick held on to his limits for a few years and then, through a syndicate which he formed, disposed of them at substantial profits. He was the only timber speculator at the Lakehead to make any real profits during the whole period from 1890 to 1920; but, in negotiating this satisfactory deal for himself, he was reported to have incurred the enmity of a former confidant, political prompter and business associate, Donald M. Hogarth. It was an enmity between the two former political leaders of Port Arthur which persisted to the ends of their active careers.

Port Arthur and the Lakehead district generally had occasion to rejoice when the Port Arthur Pulp & Paper Company began its operations. After waiting twenty years, their hope was at last realized. The small mill was enlarged shortly afterwards and its name changed to "Provincial Paper Limited." Since 1920 this pioneer pulp and paper mill has been as important to the economic welfare of the citizenry as the Pigeon River Lumber Company mill was in the preceding two decades. The future of Thunder Bay as a leading centre in pulp and paper production was now on a secure basis. Its development is to be reviewed in the next section.

The Lush Twenties: The Political Scene

Following the Armistice imposed upon Germany and its satellites on November 11, 1918, and until the summer of 1920, the North American continent went wild in an orgy of speculation and extravagance. Stocks reached dizzy heights on all the exchanges. These included that of pulp and paper, and meant continued high prices for pulpwood. Then the whole speculative structure collapsed, with prices on forest products tumbling to a depressed level. Pulpwood however, did not return to the low prices of pre-war days, but remained fairly stable from then on until the great depression of the 1930s. Fortunately, this panic was of short duration, and, by the early part of 1922, the situation was once more beginning to stabilize. President Warren Harding in his State of the Union message to the American Congress declared that economic conditions were now back to "normalcy," a term so unusual that it evoked the curiosity of the whole English-speaking world.

"Normalcy" became synonymous during the easy 1920s of an era of intense and organized lobbying for exclusive concessions or special privileges and of vast combines, international cartels, political corruption in high places, organized bootlegging, highjacking, gangsterism and speakeasies, all of which coincided with the famous Volstead Act of 1919 prohibiting the manufacture, sale and transport of liquor in the U.S. It was during this period that large and doubtful enterprises were promoted, and mergers effected and controlled by agencies outside the ownership, and sometimes even the management, of corporations. All of these events had their repercussions, some of which were of great significance to the economic life of the North American continent. The Teapot Dome scandal was the first to be smoked out; this was followed by the failure of the Chicago, Milwaukee and St. Paul Railway, and then by the collapse of the Insull Mid-West Utility Enterprises, and other great corporations.

After-dinner speakers in Canada and across the border, when addressing international gatherings, are wont to refer to our common background, our common origin, our common outlook and our common language. One of these Demosthenes actually referred to our common Bible. What these spellbinders might quite properly add to these common inheritances are common greed, common cupidity and our common urge to make a fast buck. During the same period, Canada had its Customs scandal, exposed by a committee of the House of Commons in 1926. This investigation brought out some unhealthy irregularities within the administration of the Customs Department. This led Henri Bourassa, M.P., who had just returned to Parliament after an absence of seventeen years, to state that: "What was most needed by public men in Parliament was not a keen sight, or a sensitive hearing, but a good sense of smell." A few years later the famous Beauharnois Power deal was investigated by another Commons Committee, the findings of which led one party to the "Valley of Humiliation," though only for a short time. Finally the Canadian people witnessed the collapse of the pulp and paper industry. It is against this background that the development of our Canadian pulp and paper industry and the administration of the Ontario lands and forests (one of the few remaining timber lands on the continent) is being reviewed.

It was evident, early in 1921, that the Union Government at Ottawa had lost the confidence, not only of the Canadian electorate, but of some of its own followers in Parliament. At the general election, held in November of that year, the Union forces were swept out of power. The Liberals, under the leadership of William Lyon MacKenzie King, were entrusted with the task of forming a new administration, which in turn carried on until the autumn of 1925. We have already seen, in a previous chapter, that a coalition of left- and right- wing farmers and idealistic and hard-headed labour men had succeeded in forming a government, and had been directing the political destiny of the province of Ontario since November, 1919. This strange combination of directly opposed forces did not make for a happy crew. It soon became apparent that there was irritation within the household, and that Premier Ernest C. Drury would have a difficult task in controlling this motley crew of the ship of state. Moreover — and this was the regrettable part of it — only a few of these elected representatives had had any form of training for, or experience in, the administration of public affairs.

It seems paradoxical that straight-laced political leaders rarely make good administrators. Generally speaking, if we look back over the political history of Ontario it will be readily seen that vacillation, incompetence and political corruption usually coincided with such leadership. The only explanation that can be given is that though these leaders were men of unimpeachable probity, gifted speakers, and in some cases, men of great scholarship, they were never able to estimate human character and properly assess the weaknesses and designs of their immediate followers.

In the fall of 1920, the Conservative party of Ontario was in a bad way. Largely discredited as a result of unhealthy revelations by the Royal Timber Commission in the summer of that year, they decided to choose a new leader. There were few public men among the Conservatives at that time who were able to command the confidence of the rank and file of the party, to say nothing of the electorate at large. G. Howard Ferguson, veteran member for Grenville, and formerly Minister of Lands and Forests in the Hearst administration, had the support of most timber men and lumber manufacturers, and was elected leader of the party at a convention in Toronto early in December, 1920.

As opposition leader, and more particularly on the floor of the legislative assembly, Ferguson (Fergie to his friends) revealed unusual talents, and became a formidable foe of the Drury administration. Having had considerable experience in dealing with public affairs, public men and greedy political heelers of all types, he seems to have had no doubt in his mind that the government would be an easy victim to the preying of its most trusted friends. Addressing the Conservative convention in Toronto, in 1938, he stated:

> When I became head of our party it was largely discredited with
> the people of the province. We had a comparatively small follow-
> ing in and out of the House, with little press support; and we had
> no campaign funds. Eventually I gathered enough money to-
> gether to buy a double-barreled shotgun. Then a month or two
> later I was able to procure a few shells, and made out for the
> byways and highways looking for rabbits. I saw one going
> through the fence and shot it. Then on my next trip, there were
> three or four of them by the fence, and I shot them. Finally, by the
> time the Drury administration had been in power for four years,
> there were plenty of rabbits to be shot all over the Province.

One of Ferguson's first acts when he became Premier of Ontario in June, 1923, was to appoint a commission to investigate a large provincial bond issue which had been floated through a banking house in Toronto, friendly to the Drury administration, at a price believed to be disadvantageous to the Province. A charge of conspiracy was laid against Peter Smith, a cabinet minister of the Drury administration, and Aemilius Jarvis, president of the banking firm involved. In addition to receiving prison sentences, they were jointly fined $600,000. The "rabbits" had not all been shot by Howard Ferguson, however. One, at least, had hidden in the woods, and had discovered that there were plenty of fine stands of spruce trees in Northwestern Ontario. C.W. Cox of Port Arthur, who by a stroke of good luck had emerged unscathed from the timber investigation, was rapidly developing into a tie and pulpwood operator of importance. He was in need of timber limits, or at least of a timber permit. Since he was a native of London, Ontario, it was natural that he would apply to the elected representative of that city to use his influence with the bureaucrats of the Department of Lands and Forests in obtaining the timber concession or cutting rights which he needed. Moreover, Major General D.M. Hogarth, Port Arthur's member, sat in the cool shade of the opposition, and was unable to do much for Cox at that time.

The member from London, turned lobbyist, must have taken his task rather seriously; he subsequently entered suit against C.W. Cox for services rendered in having obtained a timber limit for him. Charlie Cox, never known to part easily with his money, contested the claim. The case went to court, and Cox won the decision. Those of the electors who, by their votes and influence had elected these reformers to take control of the affairs of the government at Queen's Park, with the hope that they would soon purify public life, were badly disillusioned. They soon discovered that party labels are only skin deep, and that they do not change or influence the character of politicians; in short, it does not give them the grace to resist temptations, at least not for long.

When the general election was held in June, 1923, the farmer-labour combination was badly defeated, and the Conservative Party, with a large following, formed a government with G. Howard Ferguson as Premier. He soon came into his own. Ferguson was a man of sunny ways, an effective and exceedingly witty speaker, who quickly won the confidence of the electorate, a confidence he held until his resignation as party leader in the early 1930s. He seems to have been a political realist, an opportunist, and consistently inconsistent. There are, however, a number of notable achievements to his credit: the revitalization of the Hydro Power Commission, the amendment to the Ontario Temperance Act, the formation of a Liquor Commission, the establishment of beer parlors and, finally, the repeal of the famous Regulation 17, which had caused so many difficulties and misunderstandings in the administration of the bilingual school system, particularly as it affected the eastern part of the province. It took more than usual courage to carry out such contentious legislation. His stand on each of these public issues won for his party the approbation of the electorate. His greatest achievement however was in the field of education. To the end of his regime as leader of government and party, Ferguson reigned supreme.

The administration of lands and forests, however, was doomed to neglect. Many recommendations made by the Riddell-Latchford Commission as recently as 1922 were ignored and their report became just another document to be catalogued in the archives of the Minister of Lands and Forests and forgotten. The department once more became dominated by political patronage, political bartering and political consideration. The timber speculators and timbermen, who had been largely responsible for electing Howard Ferguson as leader of the party, who had stood by him in his days of political adversity, and who had marched with him in the wilderness were not slow to press their claims and to expect their rewards now that their party was in power. They were not disappointed.

Chapter Five

Duplicity and Intrigue

The United States had emerged from the First World War, not only as a great power, but as the richest nation on earth. Her position as a highly industrialized nation had been greatly enhanced. She had been able not only to finance her own vast war industries, and her war effort during that great conflict, but also to help finance most of her allies. She had become a chief creditor nation. It soon became evident, once the peace treaties had been signed, that New York would be displacing London as the world's financial capital. With vast surpluses of capital, American financiers and industrial leaders began to look for markets abroad, in addition to stimulating the use of new products at home. The pulp and paper manufacturers in the United States, as well as in Canada, were not lacking in vision and enterprise. With new techniques introduced in pulp manufacturing, they were able to produce many new products and by-products. Their manufacturing and technical problems were not too difficult to master. It was simply a matter of laboratory experimentation.

With the introduction in the United States of the use of wood pulp for paper making, the industry expanded with great rapidity. In 1850 there were 500 mills in the United States with a capitalization of $18,000,000 producing paper valued at $17,000,000 per annum. The value of production then rose to $48,436,935 in 1870 and to $847,279,506 in 1939, and the manufacture of pulp kept pace. The U.S. pulp and paper industry employed 137,445 people in 1939 and utilized millions of tons of wood.

The great problem with the American paper mills, and one which caused them considerable anxiety, was that production on such a scale greatly depleted their forests; they badly needed pulpwood. As a result, the invasion of Canadian timber stands was renewed with vigor. The export of Canadian pulpwood to the United States during the 1920s and 1930s reveals some startling and disturbing information, particularly when it is realized that Northwestern Ontario supplied more than half of the pulpwood exported to the United States during this period. The timber speculator with a timber option, the politician with pull, and the operators with leases of timber limits, immediately became men of great importance.

The stage was now set, and all the acts were of particular interest to the people of Northwestern Ontario and Ontario at large. In this great drama of duplicity, intrigue, lobbying for patronage, and, in some cases, highjacking for the control of Crown lands or private property, it is well to review and appraise the principal actors, and to remember that privateering, such as had been carried on prior to 1920, was no longer possible. From March of 1918 and on, all timber on newly-staked mining locations was no longer the property of the locator, but of the Crown. Timbermen, as we have seen, were a resourceful lot. Other methods of acquiring timber lands had to be used, all of which had to be within the law. Some of these methods included:

(a) A Crown permit could be obtained from the Department of Lands and Forests by which timber could be taken out. This included railway ties, piling, sawlogs, posts, poles and pulpwood. Pulpwood could no longer be exported from Crown lands. These permits could be obtained by high-pressure lobbying — in other words, by patronage.

(b) The next method would be for an operator or speculator to have a certain area of fine timber stands estimated, and by some political pressure brought upon the Minister of Lands and Forests, to have it put up for public sale, according to the Act. Usually the period of time from when it was advertised until the final date on which the bids were received, was so limited that it gave few persons a chance to submit a price; the advantage lay with the operator who had had his estimators cruise the limit. In many instances only one bid was submitted. These bids, of course, were for Crown timber, which at the time was not exportable.

(c) Still another method was that used solely by speculators who told Crown officials in Toronto that they intended to build a pulp or paper mill. They would request that certain limits be put up for auction. Again, because of limited time, there would be only one bid submitted, that of the speculators, and they obtained the timber, which, as it will be seen later, led to some profitable turnovers for the speculators.

(d) With considerable backing from pressure groups and the "boys back home," a request could be made for the opening of a township, or townships, for settlement. After completing work on their homesteads, in keeping with the interpretation of the Act, the new settlers became the owners of the land. Their wood could then be exported.

(e) A timber operator or speculator would send in a tender to the department on a timber area which had been fully advertised. He would put a bid on spruce at the prevailing price, but to enhance his bid, would quote exorbitant prices on balsam, birch and poplar, which he had no intention of cutting. With the right political pressure the limit would usually be awarded to him. This has happened in Northern and North-western Ontario.

(f) Finally, there still remained the tested method of trespassing either upon Crown property or fee simple lands, which, the records show, was indulged in quite freely.

Now, with the stage all set, and the scenery displayed in its proper alignment, we may turn to the actors. The joint control of operations held by Walter H. Russell of both Russell Timber and the Newaygo Company terminated in 1921, when the interests of the latter were purchased by the Consolidated Water Power and Paper Company of Wisconsin Rapids, and George Schneider of Appleton was appointed manager. Russell Timber, having been badly mauled as a result of their encounter with the two learned judges of the Royal Timber Commission in June of 1920, were fighting for survival; but they still had a considerable number of patented timber and mining lots under their ownership, in addition to other private property of fee simple lands, from which they could continue their operations. For a few years, at least, they were still to be active in the exporting of pulpwood to American mills. If it is true that Walter Russell held the centre of the stage in the second decade of the present century, his role in the early 1920s was secondary. He eventually was relegated to a minor position.

This much can be said of Walter Russell in his heyday: he was a good spender, lived in semi-regal fashion, had a magnificent home and beautiful gardens, and was a lover of flowers. He had a fine yacht, with which he was most hospitable to his friends. A trip on the yacht "Pom-Pom" was an event in any man's life. Walter Russell was fond of good food. He was a gourmet. When taking a party out on the lake, his favorite act was to don the chef's cap and jacket and cook the meals. He was particularly efficient in preparing stewed chicken with dumplings. On one occasion the pilot missed the channel to one of the islands where the Russell Timber Company was operating a camp, and ran aground on a sand bar. In rocking the boat to get it over the bar, the whole stew was upset on the floor. The guests went without a most tempting dish. A good organizer, Russell was an early president of the Lakehead Rotary Club, and eventually president of the Canadian Pulpwood Association. He took an active part in many activities, including the church and the social life of his adopted city. His final disappearance from the stage deprived the pulp and paper industry of a cultured and colourful figure.

From 1919 until June, 1923, with a farmer-labour combination in power at Toronto, the followers of the old Conservative party which had exercised such authority from 1905 to 1919, did not, naturally, have much influence with the reformers who were now in charge of the ship of state. However, Beniah Bowman, Minister of Lands and Forests, did hand out a few lush concessions as already related in the Cox incident. It was not what may be termed a period of progress so far as the industry was concerned in and about the Lakehead. Most of the operators were already marking time.

C.W. Cox, however, was forging rapidly ahead. He was taking out more pilings, pulpwood and railway ties from some of his mining timber lots which he had staked a few years previously along the south shore of Sheshbeeb Bay, on the Black Bay Peninsula, by some trespassing, and from the timber limit which he had earlier obtained from the Drury government. He also continued to trespass on crown and other land. He had been able, by temperament, to adapt himself to his environment. He set his sails to all the winds. It could be said that he slept with all the parties. By 1926, he secured the exclusive rights on all the spruce from the Indian reserve at Long Lac, from the federal government. This wood was exportable. The demand being good for spruce pulpwood, he sold his entire timber rights on the reservation to the Detroit Sulphite Company. It was reported that he had made a profit on this transaction of $60,000. He was in the money, built himself a magnificent home, and took a trip with his wife to Europe. He then decided to enter politics. He had already been president of the Chamber of Commerce, and was elected as an alderman. In 1934 he was chosen mayor of Port Arthur, and served as chief magistrate for a period of fifteen years. During the same year he was elected as a member of the Ontario legislature as a Liberal. With all the fervor of the convert (he had been a Conservative) he became a staunch supporter of the Liberal policies propounded by "Mitch" Hepburn, the Huey P. Long of Ontario. He was made a member of the cabinet without portfolio, but did not seem to be able to get along with Peter Heenan, Minister of Lands and Forests. In the reorganization of the cabinet in 1937, he was left on the outside. For a reason that will be examined later, C.W. Cox parted company with the government on the policy of exporting pulpwood which he himself had enunciated, and for which he had obtained the support of all his fellow timbermen at the Lakehead. It is said that at one time he flirted with the Co-operative Commonwealth Federation. Eventually he lost the Liberal nomination in Port Arthur to Major Bert Styffe. He once described his role in the legislature, as "the voice of one crying in the wilderness."

In the early 1920s, the Fort William Paper Company pulpmill was built on the Mission River at Fort William. It was in immediate need of pulpwood, and a contract was given to Joe Greer, a partner in the firm of Greer Brothers, who conducted heavy tie operations during the Grand Trunk Pacific Railway construction period. He had a monopoly on all the wood required, and for the next few years seems to have prospered, and to have regained, to some extent, the status that his firm had once enjoyed as railway tie contractors. But Joe Greer was shortly to disappear from the scene.

C.H. Greer had operated along the P.A.D. & W., as a railway tie contractor and a logging operator. He then went into the hotel business, and for many years was content to look after the management of his investments. He fitted in well as a Boniface. But, with the enactment of the Ontario Temperance Act, his hotel operations became unprofitable, and he launched out once more as a timber operator. Beginning modestly in the early twenties, he quickly became the largest railway tie contractor in the district, taking out between 500,000 and 600,000 railway ties a year. He was unquestionably a good timberman, and got along well with the bushmen. He, like Cox, was able to take a trip to Europe with his wife. The demand for ties being over by the mid thirties, Greer went into mining and other ventures, with which he was unacquainted. These speculations proved disastrous. His operations had been largely confined to ties. Familiar with all the short cuts of his time, he could hold his own against any competitor. And he did not worry too much about trespassing over the lines.

William Scott, already past the meridian, and never robust, organized the Scott Timber Company. Later he formed a sort of partnership with Oliver Schram as dealers in railway ties and pulpwood. After a few years Schram pulled out leaving Scott to continue until the late thirties. Hampered by insufficient capital, having been one of the victims of the real estate bust of 1913-14, and not in the best of health, he did not rise to any great heights. His various transactions in pulpwood and railway ties were not too profitable, particularly during the years of the depression.

Thomas Falls, a railway tie contractor during an early period, obtained concessions from the Crown, and took out some substantial pulpwood contracts during the 1920s. Actually, he was a railway tie man; it was in this field that he knew his ground best. He was of a flexible character, and adapted himself to the environment of the time; he seems to have got along without too much friction, or too much remorse, so far as wandering or trespassing was concerned. A rather good-looking man, and most affable, he had many friends; a true pioneer logger in the district.

In 1920 the Detroit Sulphite Company of Detroit, Michigan, which had been obtaining pulpwood during the previous thirty years from local timber contractors at the Lakehead, decided to open up an office and yard at the Lakehead, with a woods supervisor of their own. Shortly afterwards, Walter Laubengayer took up residence at the Lakehead. The company obtained water frontage on the right bank of the Kaministiquia River, near the old Grand Trunk swing bridge, opposite the N.M. Paterson grain elevator. There they had a large pulpwood yard, with some rather novel equipment for loading the wood from the yard into the boats. Until the mid-1940s, they still had their busy office in Port Arthur to handle exclusively the exportation of pulpwood to their own mills. They later maintained an office at Sault Ste. Marie, Ontario.

The Central Contracting Company, who, as already recorded, had gone into liquidation early in the 1920s, completely disappeared from the scene. Their equipment was taken over by the Hammermill Paper Company, who decided to carry on their own operations, with Malcolm Cochrane as general manager of their woods operations and woods purchases. From then on they became important in the exportation of pulpwood to their mill at Erie, Pennsylvania. This firm, along with the Port Huron Sulphite Company and the Detroit Sulphite Company, had been buying pulpwood continuously from this district since the late 1890s. Their operations, by the late 1950s, were quite extensive, particularly in the eastern end of the district of Thunder Bay, and in the districts of Algoma and Temiskaming.

In the 1920s, the Newaygo Timber Company, under the supervision of their new manager, George Schneider, was making steady progress, not only in their own operation, but in obtaining vast timber areas. Schneider seems to have been well equipped by training, disposition and outlook to fit in with the picture of the decade. He had had experience as a salesman in the American mid-west for one of the large lumber manufacturers of the west coast. He was flexible, and a good organizer, particularly of political support. He seems to have been quite aware that elections, even provincial elections, are not won with prayers, and that timber concessions are more easily obtained from a government just prior to a general election than after. To operate the machinery of party politics under our system of government, the parties must have campaign funds. This necessity is as old as our system. These funds are not usually obtained from idealists, but rather from interested individuals, groups of individuals, or from companies who are likely to be recompensed in concessions handed out to them for whatever generosity they have displayed to their party before an election. This has been quite common in the timber industry. George Schneider seems to have possessed a knack of getting results, even if the means to achieve that end, as reports would indicate, were costly to his company, but rewarding to the brass hats of the political party which held power. He disappeared from the scene altogether in the early 1930s, and was succeeded as secretary and manager by H.S. Mosher, who consolidated the various interests and subsidiaries of the company under one head. By 1960, they had become one of the largest exporting firms in all Northwestern Ontario, with an average of more than 75,000 cords a year.

By the late 1920s, Oscar Styffe had formed an organization of his own (later operating under the name of Oscar Styffe Ltd.). A native of Norway, he had worked many years in the northern woods of Wisconsin and Minnesota before coming to Port Arthur. A man of high intelligence, he was a good timber man, and in the matter of a few years had worked up a large volume business for himself. Towing his pulpwood across the lake, he lost one boom containing a large amount of pulpwood, some of which was recovered at a heavy cost. This loss hampered his operations for the next few years, particularly during the Great Depression, but subsequently he began to expand again, and his sons carried on the business founded by their father. Their volume was substantially large enough to make them a factor at the Canadian Lakehead.

Another timber operator of the late twenties was Charles W. Gardner, formerly woods superintendent of the Provincial Paper Company. Charlie Mellor, a pioneer mill owner along the P.A.D. & W., had obtained a timber limit and to exploit it, a syndicate was formed by himself, Charles Gardner and Arthur G. Pounsford (also of Provincial Paper). Pounsford withdrew at the end of one year, and subsequently Charlie Gardner bought out the Mellor interest in the limit, and continued until his death.

The Rise of E.E. Johnson

A new contender for space under the sun and his share of the forest wealth of Northwestern Ontario entered the scene in the early 1920s. E.E. Johnson, who had wound up the assets of the old Pigeon River Lumber Company in 1919, entered on the scene as an independent operator. With a mind as quick as a flash, this American-born descendant of the Vikings had been reared and educated in Wisconsin where he had received his early training. A law graduate of the university of his native state, he seems to have possessed all the vision of Lief Erickson, the energy of Thorfin Karlsefin, and the enterprise of Lief the Lucky; E.E. Johnson had inherited all the restlessness and some of the ruthlessness of his remote ancestors.

Johnson's first venture on his own was as a pulpwood contractor. Having obtained a contract from the newly-organized Fort William Paper Company, he took out 16,000 cords of pulpwood from a timber location on the inside of the Black Bay Peninsula, from which the Pigeon River Lumber Company had taken out sawlogs for two winters. These operations were apparently highly profitable. Having proved to himself his ability to carry out fairly large undertakings, Johnson immediately launched into other operations. He took out another rather heavy cut of pulpwood; but since it was from Crown lands, and not exportable, he traded his pulpwood with the Provincial Paper Company for white pine sawlogs which this company had been forced to cut on their Sibley limits. The large sawmill of the Pigeon River Lumber Company had been closed down, and to put it in a state of repair would have been too costly. To saw logs into lumber, Johnson decided to erect a small sawmill alongside the larger one, exclusively operated by electric motors. He may not have known it at that time, but he undoubtedly pioneered electrified sawmill operations in the District of Thunder Bay. Although the price of lumber was not high at the time, he was able to dispose of his entire cut at a good profit.

Greatly stimulated by his ready success, Johnson now began to look for timber and hired "Tar-Paper" Pete Grant, former superintendent of woods operation for the old Pigeon River Lumber Company, as timber cruiser. For the next few years this veteran of logging operations was kept exceedingly busy exploring and estimating timber stands all around the district. He died in Johnson's employ. Oscar Lehtinen, an experienced logging executive, was made general superintendent and a partner in the firm. Eddie Johnson challenged the authority and the rights of the Pigeon River Lumber Company, with which he had been formerly associated as secretary-treasurer, over the Arrow and Pigeon Rivers. He and Charlie Cox had been cutting pulpwood and railway ties along the banks of these rivers. Johnson had a partnership with W.J. Cochrane, formerly a timber cruiser of the Pigeon River Lumber Company, to take out some 25,000 cords of pulpwood on Indian land on the American side of Pigeon River. Cox had been cutting some 40,000 railway ties on the upper stretches of the Arrow River. They jointly drove their pulpwood and their ties together to the mouth of the river, disregarding the charter rights of the two subsidiaries of the Pigeon River Lumber Company (the Arrow River Slide and Boom Company, and the Pigeon River Tributaries Slide and Boom Company) to collect dues for river improvements. Their audacity did not go unchallenged. Johnson's and Cox's timber in booms was seized at the mouth of the Pigeon River. This led to a suit which was fought in the district courts of both Thunder Bay, and the state of Minnesota, and both courts recognized the authority and rights of the Pigeon River Lumber Company and their subsidiaries. This decision was further confirmed by the Supreme Court of the United States, and the Supreme Court of Canada.

Walter H. Russell, who still had considerable timber lands in fee simple, had organized the Arrow Land and Logging Company in which E.E. Johnson and Ben Alexander of Wausau, Wisconsin, held a two-thirds interest. It soon became evident that Eddie Johnson was the mastermind of the enterprise. They operated a sawmill for some years near Hearst where they held a sizeable limit of fine spruce trees suitable for sawlogs. A.L. Johnson, a brother (later Colonel A.L. Johnson), was appointed manager. It seems to have been a successful operation.

It was in the 1920s that Johnson took out a large contract for pilings, in partnership with Charlie H. Greer, for delivery to some of the large grain elevators then under construction on the Port Arthur harbourfront. Meanwhile, Johnson had acquired some small diesel operated tugs for hauling supplies to his camps, preparing booms and towing pulpwood. He was forging spectacularly to the front as an operator, and with his brother, began to devise more modern methods in timber cruising. He brought in the first snowmobile ever used at the Lakehead, a conveyance in which they were able to fly over the ice of the bays and lakes bordering the North Shore. Eventually his firm bought a large aeroplane, in partnership with C.H. Greer, which was piloted by Al Cheesman. It is believed to be the first plane to be used under private ownership for timber cruising in Northwestern Ontario. Ever innovative, the Johnsons transported their pulpwood to dock at Heron Bay by an ingenious system of two jack ladders and a continuous flume to which the water was pumped and the wood floated by gravity to a storage boom, or right into the vessel.

From his accumulated profits, E.E. Johnson organized the Pigeon Timber Company Ltd. The name had no significance so far as Pigeon River was concerned, but he was quick to realize when he came to Port Arthur, that the old firm, the Pigeon River Lumber Company, enjoyed high prestige in the District. Ben Alexander of Wausau, a former partner of the old Pigeon River Lumber Company, along with some of his associates, became interested in this new venture and contributed up to forty percent of its capitalization. Their share was subsequently purchased by E.W. Backus, founder of the Backus Brooks Company of Northern Minnesota fame, and subsequently the Ontario and Minnesota Power and Paper Company of Fort Frances and Kenora, and finally the Great Lakes Paper Company of Fort William. With Backus as a partner, Johnson exercised a monopoly on all the pulpwood requirements of the Great Lakes Paper Company. From then on, his rise to fortune was easy. He moved from conquest to conquest. In addition to handling his tremendous contracts, he was also engaged in exporting pulpwood to American mills. His problem, like that of other pulpwood operators, was the acquisition of timber lands and timber limits. A subtle lobbyist, and a cold-blooded calculator, he was a tough negotiator. By a strange contradiction in his character, he could at times be generous, and as circumstances dictated, he could be charming. He could cajole, flatter, even bend a knee to, or threaten, his political friends at the Lakehead, or thunder at the seat of the mighty in Queen's Park. His political campaign contributions were many and generous, but always handed out with the expectation of substantial rewards in timber concessions.

Johnson bought the Bowlker Farm on the banks of the Kaministiquia River, some few miles west of Fort William, on which he established a fur farm. His mink ranch became one of the best known on the continent. He modernized all the farming equipment, and made the soil of this large area highly productive. Eventually in 1940, he organized and

successfully launched the greatest of all his enterprises, the Great Lakes
Lumber and Shipping Ltd. Two years previously he had accumulated a
strange fleet of overage tugs and barges for the transportation of his
pulpwood to American mills; hence, the name of his company. His
large sawmill on the Mission River flats at Fort William was electrified,
and had the most modern equipment of any of the mills in Northern or
Northwestern Ontario.

This was not all simply accomplished by a stroke of the pen, or by
the raising of capital. A rugged individualist, and out of step with the
trend of the time, Johnson met with serious labour difficulties in his
logging operations and at his mill. He was obliged, in time, to recognize
that labour was a force to reckon with — that they had certain rights.
But it was a long struggle before he was willing to admit this in his own
mind. Eventually he became an ardent exponent of organized labour.
His earlier methods of pulpwood operations were not at all according to
the tenets of the Bible. His sins of trespassing, like those of his contem-
poraries, were quite common, and if they became unforgivable, he usually succeeded in inducing the Department of Lands and Forests, through
his political friends, to grant him areas sufficiently large to cover the
timber that was discovered to have been "mistakenly" taken out from
Crown property. According to some reports, his woods superintendents
had not been too particular about separating the burned wood from the
green wood. Immediately after a fire, there is considerable timber which
can be used if it is cut while still green, although it may have been badly
scorched. The Crown sells that wood at distress prices. It is therefore
a temptation for an operator to mix the two together. None at the Pigeon
Timber Company lost much sleep keeping the two separated. Taken as
a whole, despite the questionable methods that he chose to use at times
in his operations, E.E. Johnson has undoubtedly made distinct contributions to the economic life of the District of Thunder Bay.

Eventually, with changes at the political front in Toronto, a new
group of politicians arose who were less susceptible to Johnson's per-
suasive powers. Moreover, the reallocation of timber limits to large
pulp and paper corporations left Eddie Johnson more or less restricted
in the concessions which he obtained. He was, by the late 1940s, at
odds with both the pulp and paper mill interests, and the officials of the
Department of Lands and Forests. For a few years he spent considerable sums of money in fighting a losing battle for sawlogs, which were
by then his main interest, in order to keep his sawmill going. He re-

ferred to this struggle as a "holy war." Eventually his large sawmill on the delta of the Kaministiquia River had to close down. Although a shrewd business executive, he had been favoured a great deal by luck and good timing. By this time he had accumulated millions. Always a dynamo of energy he died suddenly. Though he achieved much and accumulated millions of dollars personally, Johnson's contributions to the industry were not lasting. His era was rapidly passing.

Donald A. Clark, son of A.L. Clark, pioneer tie contractor, came to the Lakehead in 1906, and was one of the young actors attempting to crash into this great play of the 1920s. Eventually he was to carve himself a great career and a substantial business out of the forests of Northwestern Ontario. His beginning was modest, but he was endowed with a winning personality, and exceptional ability, which ensured that his rise was to be consistent. Like most men destined to play leading parts, he had to overcome some pitfalls. But he had abundant faith in himself, and in the future, and being exceedingly resourceful, he was able to overcome these obstacles.

Don Clark was born in Dresden, Ontario, but moved with his father to Port Arthur when he was quite young. The blood of the pioneer frontiersman was in his veins. His father was a timberman, and his grandfather, although a farmer, had conducted logging operations during the winter months. Quitting public school before finishing the eighth grade, he was at work in his father's operations at the age of 15. A year later he was in charge of driving operations along the Matawin River. He remained with his father, as a partner, until the latter retired in 1920. He then took over the family firm.

Clark's first contract of importance was with the Kaministiquia Pulp & Paper Company, which had just built a small pulp mill known as the Little Mill located on the west side of Current River near the lakefront. This plant had barely been completed before it went into receivership, and D.A. Clark was left with 10,000 cords of wood on his hands. He succeeded in selling the wood to the Central Paper Company of Muskegon, Michigan, which soon became a steady customer. A few years later, in 1926, he took out a heavy cut for the Nipigon Corporation at Nipigon, and in that operation he was most successful. He was then on his way to further expanding his operations.

During the Great Depression, like most of the woods operators of that period, he found it hard going with limited turnovers and permanent overhead. A few years later, however, Clark was fast recovering lost ground, and by the 1940s, had developed into one of the largest operators in the district. He took a leading part among the great actors of a new drama now being played by the "haves" and the "have-nots." During 1946-47, his companies actually took out 200,000 railway ties, 140,000 cords of pulpwood and 50,000 hydro poles, representing a value of $3,500,000. They carried on operations at Nipigon, Beardmore, Wabinash, Clarkdon, Kowkash, Sunstrom and Long Lac, employing a total of 1,700 bushmen. In addition to operating under the name of D.A. Clark & Company, he was also the controlling factor in the Nipigon Lake Timber Company Ltd., of which he was the president. He was also vice-president of the Alexander Clark Timber Company, which was organized in 1945, and had extensive operations in the Sturgeon Lake District near Sioux Lookout. Don Clark's organizations took out pulpwood for shipments to the following firms: the Hammermill Paper Company of Erie, Pennsylvania; the Port Huron Sulphite Company of Port Huron, Michigan; the Neekosa-Edwards Paper Company of Port Edwards Wisconsin; the Filer Fiber Company of Manistee, Michigan; and the Newaygo Timber Company of Port Arthur.

Don A. Clark had the great advantage of appearing on the scene after 1920. He had not, therefore, been involved in the period of pirateering. In all of his transactions with his clients, he was always above board. The greatest tribute that can be paid to him is that they renewed their contracts year by year. His relations with his many employees were usually harmonious. His greatest strength was his ability to get along with his men. A born organizer, he was able to develop a general staff that commanded respect and obtained results. Of course he had to operate in a period and against competitors who had little regard for the lines allotted to their various operations by the district foresters. To trespass on Crown reserves was just as good sport in the 1920s as it had been before.

Trespassing in cutting pulpwood and ties became so prevalent in the late 1920s, in the District of Thunder Bay, that the heads of the Department of Lands and Forests in Toronto decided that it could no longer be ignored. Accordingly, a high-ranking official was sent to Port Arthur with full authority to straighten out the Timber Wolves, make out an estimate of what timber had been cut from Crown lands, and assess all of those guilty of trespassing. From a report of the time, it seems quite evident that practically all of them had to settle on a basis

The Graham and Horne
Company (above) was one
of the earliest to establish a
sawmill operation at the
Lakehead. Their mill was
located on the north bank of
the Kam River in Fort
William. (Thunder Bay
Historical Museum Society,
972.2.359)

Charles W. Cox was one of
the most notorious timber
operators in Northwestern
Ontario. A perennial thorn
in the side of politicians in
Toronto, the populist Cox
remained a favourite of
voters in Port Arthur for
several decades. (National
Archives of Canada,
PA53848)

A typical pile driver in operation, this one at Nipigon in 1942. Photo by E.C. Everett (Thunder Bay Historical Museum Society, 974.2.99B)

One of Charlie Cox's hauling operations and camps (below), probably in the 1920s or 1930s. Photo by Lovelady (Thunder Bay Historical Museum Society, 984.53.1154E)

A view of the dining hall and kitchen staff at Red Sucker camp in the 1930s or 1940s. If a camp was to operate efficiently, it was extremely important to have good cooks. Photo by E.C. Everett (Thunder Bay Historical Museum Society, 974.2.80F)

A log jammer in operation. This was part of the Newago Timber Company's operations near Nipigon in February 1944. Photo by E.C. Everett (Thunder Bay Historical Museum Society, 974.2.80V)

Timber pilings being laid for the foundations of one of Port Arthur's terminal elevators in the early years of the twentieth century. Photo by Lovelady (Thunder Bay Historical Museum Society, 984.53.23A)

Inside the Great Lakes Lumber and Shipping Company's sawmill in Fort William (above), c.1948. Daily output from this modern mill averaged 500,000 board feet of lumber. Photo by McKim Advertising (Thunder Bay Historical Museum Society, 976.4.25A)

Don Clark (left) with Lorne Sharp. Clark was one of the most successful independent timber operators in Northwestern Ontario. (Thunder Bay Historical Museum Society, uncatalogued)

A typical Northwestern Ontario water sled being pulled by a tractor — part of the Newago Timber Company's operations in 1944. Photo by E.C. Everett (Thunder Bay Historical Museum Society, 974.2.81L)

One of the Great Lakes Paper Company's snow rutters at Valora, 1942. Photo by Steve Zuffa (Thunder Bay Historical Museum Society, 978.98.2Y)

Breaking a log jam could be both strenuous and dangerous. But it was also a very delicate operation that required a great deal of skill. Photo by Herb Nott & Co. Date unknown. (Thunder Bay Historical Museum Society, 975.75.22)

Much early logging in Northwestern Ontario occurred along the Pigeon River. Here logs are being diverted to a chute bypassing High Falls. Photo by J.F. Cooke (Thunder Bay Historical Museum Society, 983.86.101)

This early view, probably along the Port Arthur, Duluth and Western Railway, shows a primitive loading apparatus and what appears to be a rather precarious load, probably a "brag load". (Thunder Bay Historical Museum Society, 981.39.97B)

Executives of the Great Lakes Lumber and Shipping Company, l-r: W.E. Hunt (Vice President), E.E. Johnson (President); A.B. Ellingson (mill manager). Johnson's company was perhaps the largest sawmill to operate at the Lakehead. Photo by McKim Advertising (Thunder Bay Historical Museum Society, 976.4.16B)

A typical Northwestern Ontario logging camp of the late nineteenth century. (Thunder Bay Historical Museum Society, 984.70.10A)

This photo shows the style of costume and equipment used by nineteenth-century lumberjacks. (Thunder Bay Historical Museum Society, 984.70.9)

3,000 cords of pulpwood being towed 23 miles to the Dryden Paper Company's mill from its divisional depot. Photo by Paul E. Lambert (Thunder Bay Historical Museum Society, 987.31.49)

The Pigeon River Lumber Company's bush operations, c. 1908. Photo by Forde (Thunder Bay Historical Museum Society, 979.102.18D)

The giant Port Arthur sawmill of the Pigeon River Lumber Company, c. 1912. (Thunder Bay Historical Museum Society, 979.102.18C)

Edward's sawmill in Sellars, c. 1907, is typical of the small family-run operations throughout the region. (Thunder Bay Historical Museum Society, 983.70.11)

The Provincial Paper Company and, in the background, the Thunder Bay Paper Company mills, as seen in c.1960. Both mills were constructed in the 1920s. (Thunder Bay Historical Museum Society, 973.28.10A)

After dinner recreation at the logging camp, featuring a little fiddle music, a few rounds of with the boxing gloves, and a friendly game of cards. The date and exact location of this photo are unknown. (Thunder Bay Historical Museum Society, 972.2.144)

The Tom Falls Timber Company's snowbound camp at Shebandowan, Ontario, date unknown. (Thunder Bay Historical Museum Society, 984.101.1A)

The Abitibi Power and Paper Company clearly spruced up this bunkhouse for the photographer. (Thunder Bay Historical Museum Society, 976.4.10E)

Workers for the Detroit Sulphite Pulp and Paper Company load pulpwood at Nipigon for export to American markets. Photo by Lovelady (Thunder Bay Historical Museum Society, 984.53.1320)

that brought considerable additional revenue to the provincial treasurer.

There were, of course, other contractors who played no inconsiderable part in taking out pulpwood in the 1920s, namely, John Kallio, Sam Hill and A.L. Allard. There were also a number of pioneer Finns in the district who took up farmland during that period, and were subcontractors to the larger organizations. In time, these Finnish timbermen established substantial farm holdings where they produced hay and roots for which they found a ready and profitable market with timber operators.

Political Influence

The history of the operators during the abundant twenties having been reviewed, it may be well to examine the political setup at the Lakehead during that period. In 1919, General Donald M. Hogarth was re-elected to represent the district on the Conservative ticket. Don Hogarth, as he was known to his friends, was a native of Mattawa on the Ottawa River, where he received his primary education. He then moved to Sault Ste. Marie, and joined the various Clergue organizations with which he was associated for a few years as purchasing agent. J.J. Carrick, then a salesman for the Imperial Oil Company, came to Port Arthur in 1904 to open up a real estate office, and brought Hogarth with him. The two were associated intimately until their breakup in the early twenties over the Black Sturgeon and Pic River timber deal. General Hogarth was a gifted organizer, and a born politician. He had mooted J.J. Carrick's nomination in 1908 for the legislature against George Mooring, a lifelong Conservative. Carrick won decisively, and was M.P.P. for Port Arthur until 1911. Again Don Hogarth organized the Conservatives of the district to have Carrick nominated for the federal house. It was undoubtedly one of the best organized campaigns in the history of the district. Because of the vastness of the constituency, the election had been deferred for two weeks after the general election. The redoubtable James Conmee, a former mayor of Port Arthur, a member of the legislative assembly, and finally of the House of Commons, capitulated before going to the poll. Carrick thus received an acclamation.

Some two months after the federal election, Sir James Whitney, with declining health and vitality, decided to test public opinion in the province. At the general provincial election which ensued, Hogarth won a bitterly contested campaign against Frank H. Keefer, veteran attorney-at-law in Port Arthur, and easily defeated the Labour candidate. In 1914 Hogarth was re-elected to the legislature, and despite the debacle of the Conservative party, won again in 1919. In 1923 he supported the nomination of Frank H. Keefer. In the following election, however, he defeated Keefer, and was a member for Port Arthur until he withdrew altogether from politics early in 1934. But whether he was a member of the government, sitting with the opposition, or playing the role of a private citizen between elections, Don Hogarth was a tremendous force with party leaders in Toronto. His influence in obtaining special concessions and privileges was readily recognized by his party friends.

J.J. Carrick, on the other hand, was practically negligible as a political force. Though brought up as a Liberal, he became a Conservative candidate, and was elected to the legislature in 1908, and then to the federal house in 1911. Failing to obtain the nomination in November of 1917 as a Unionist candidate, he ran as an independent Conservative in 1925, and again in 1926. In 1930, he finally returned to his first love, and actually accepted the nomination to run as a Liberal. Having lost the confidence of party leaders early in this period, he never again became a political factor. This power was invested in D.M. Hogarth.

In and about Port Arthur there were two leading businessmen who had early recognized the possibility of enlisting their services under the Hogarth banner. The first one was a former Liberal, W.T. McEachern, who operated a drug store in Port Arthur. A rather mysterious figure, he played the role of the "Grey Eminence," manoeuvring behind the scenes. Don Hogarth did not require a political mentor, but in this particular case the "Grey Eminence" was undoubtedly his prompter.

The other influential political figure in Hogarth's entourage was Colonel J.A. Little, who came to Port Arthur in the early part of the century to take over the management of Molson's Bank, which was later absorbed by the Bank of Montreal. Little was not the type of bank manager to remain long contented with his daily duties, yearly salary, and the assurance of a pension in his declining years. Restless, proud and ambitious, he began, shortly after his arrival at the Lakehead, to carve an alternate career for himself.

During the summer of 1906 he joined the ranks of the 96th Lake Superior Regiment, then in the process of reorganization. Endowed with a distinguished bearing, and possessing a martial and commanding appearance, he was appointed a major of the regiment for Port Arthur.

J.J. Carrick, who delighted in poking fun at the leading personalities at the Lakehead in his real estate advertisements, always referred to J.A. Little as J. Army Little. In 1914, Little was gazetted as Lieutenant Colonel of the regiment, but did not see overseas service. It was held that his presence was more important at the Lakehead, guarding the many grain storage elevators against sabotage.

Colonel Little was unquestionably an able executive; aggressive, methodical, and shrewd. He fitted in well in that period. As bank manager, he was in a position to grant or recommend loans to the customers of the bank, and occasionally obtain a silent partnership with some of the individuals or firms to whom loans were advanced, particularly if he saw a chance to increase his earnings. Not all his investments or partnerships became profitable. On the whole, however, he steadily rose to the status of a successful financier and enterpriser.

An ardent party man, of strong Conservative tendencies, Little was familiar with the inner mechanisms of political organizations. He was well aware that elections were not won with prayers. He stood high with the party brass at Ottawa from 1911 until 1922. Except for four years during the period of the Drury administration, he exercised great influence over the sultans of Queens Park. His greatest influence was reported to have been with the Ferguson administration. Indeed, he was one of the principal organizers behind Howard Ferguson's successful bid for the leadership of the provincial Conservative party.

By the end of the First World War, Little had his fingers in so many pies at the Canadian Lakehead, and on so many payrolls, that J.P. Mooney, the wit of Port Arthur, and a former associate of J.J. Carrick, referred to the Colonel as the hero of a hundred payrolls. Little was by now a man of great influence and substance in Northwestern Ontario, with interests in lumber manufacturing and railway tie contracting. He also held exclusive commercial fishing rights on Lake Nipigon, and owned the Empire Hotel in Fort William, which he had been able to buy at greatly depressed value after the Ontario Temperance Act became law in 1916. He immediately leased it to the Royal Canadian Mounted Police. This last deal was commented upon by F.B. Allen, pioneer publisher at the Lakehead, who wrote a pun about it: "As Jimmy would say, don't put all your eggs in the same basket." Little then took over the complete ownership of the Thunder Bay Harbor Improvement Company Ltd. which, under his able management, prospered until his death in 1931. He continued to take an active part in provincial politics. He was a key man in obtaining timber concessions for himself, or for other operators, in return for campaign contributions. Little was a social and political cynic. When negotiating political deals of rather nebulous nature, he would quote (according to reports from his close associates) from Proverbs 22: 1: "A good name is rather to be chosen than great riches. And loving favour rather than silver or gold."

J.A. Little was unquestionably one of the leading and most forceful personalities of Northwestern Ontario from the time he arrived at the Lakehead until his death. He can be regarded as a veteran soldier, banker, financier, lumber and tie manufacturer, and one of the Timber Wolves of his period.

Chapter Six

The Rise and Slump of the Pulp and Paper Industry

P ulp and paper manufacturing in Canada developed rather slowly. It was a relatively unimportant industry until the last two decades of the nineteenth century. The first mill to manufacture paper in Canada was established at St. Andrews East, Quebec, on the North River near its confluence with the Ottawa River, a short distance from the Lake of Two Mountains. This small mill, which utilized the water of the rapids going through the village, was erected in 1803; it manufactured paper from rags, the common practice at the time. Three more mills were constructed during the next twenty years, one at Portneuf, Quebec, another at Hamilton, Ontario, and the third at Halifax, Nova Scotia. By 1851, ten such mills were in operation.

It is claimed that Alexander Buntin installed the first wood grinding machine to be operated on the North American continent at Windsor Mills, Quebec, in 1866. The total number of pulp and paper mills in operation by 1881 was 36 with a total of 1,598 employees, a payroll of $460,476 and an output valued at approximately $2,500,000. This was the origin of our pulp and paper industry in Canada.

During the next decade, the use of wood pulp in paper making was more extensively developed. At the beginning of the twentieth century, the value of the output of the industry exceeded $8,000,000 per annum. From then until the late 1920s production underwent a spectacular growth. It is stated that it is an ill wind that blows no one good. The forest reserves of the United States were being depleted whilst the demand for paper kept increasing, forcing American paper manufacturers to turn to Canada for pulpwood and pulp products. Thus it was that Canadian pulp producing facilities began to expand so rapidly. Canadian sentiment in the late 1890s grew increasingly strong against the exportation of raw pulpwood as well as sawlogs to American mills, and provincial governments enacted legislation prohibiting the export of pulpwood cut from Crown lands. To relieve the situation, the United States in 1909 lowered its tariff on newsprint and completely removed it four years later. Thus stimulated by wider markets, the production of pulp and paper in 1919 had reached a total value of more than $100 million. By the end of the third decade of the century, however, it was evident that the market was becoming over-saturated. In the early 1930s, a large number of companies were facing serious financial difficulties and some of the larger organizations went into receivership and trustee-ship. Only by a rigid system of economical operations and refinancing, was the industry as a whole able to emerge from the slump. Since then, however, growth has been phenomenal spurred on by new uses for pulp and paper products and by steady demand for newsprint. By 1959 there were 112 pulp and paper mills operating in nine provinces (a $3 billion investment) with a combined annual production of 10,733,000 tons valued at over $1.5 billion. The industry employed 65,985 people in the mills and 293,000 bushmen, with a combined annual payroll of $518 million. The most extensive forest resources are in Quebec with 35% of the total accessible forests followed by Ontario and then B.C. The pulp and paper industry is the largest tenant of provincial forests holding lease on an area of nearly 130,000 square miles and owning some timber lands outright.

The pulp and paper industry had a humble origin in North and Northwestern Ontario. The first mill to be built was erected in Sturgeon Falls in 1894 by the firm of Paget, Heat & Company of Huntsville, Ontario. It was a small groundwood mill. They acquired the power rights on the Sturgeon River from Martin Russell of Renfrew, and entered into an agreement with the small town of Sturgeon Falls, by which they received a bonus of $7,000 to assist in financing the enterprise. They operated only a short time before they encountered financial difficulties. The plant was then sold out to an English company which obtained a charter to operate the mill under the name of the Sturgeon Falls Pulp and Paper Company Ltd. The firm brought new life to the enterprise by expending considerable capital in modernizing the equipment. This industry was of such importance to the small community of Sturgeon Falls that it was exempted from taxation.

The mill changed ownership frequently over the next few decades. In 1903 it was taken over by the Imperial Paper Mills of London, England. The new owners manufactured paper until 1911, when the plant was sold out to the Ontario Pulp Company. By this time it represented an investment of $1.75 million. By the following year, the company was in liquidation and its assets were purchased by the Spanish River Pulp and Paper Mills Ltd. It, in turn, was taken over in 1927 by the

Abitibi Power & Paper Company Ltd., who operated it for a short time, manufacturing groundwood pulp, sulphite pulp and newsprint. In 1946 the company began a program of rehabilitation, and when operations were again resumed by the fall of 1947, the Sturgeon Falls mill produced a daily capacity of 100 tons of corrugated and hard board semichemical pulp, thus utilizing almost every variety of wood species found in that district. Their own power plant at Sturgeon Falls developed 7,500 electric horsepower.

The second mill in Northern Ontario was built at Sault Ste. Marie in 1895 by the Clergue interests. It originated as a water power investment headed by Francis H. Clergue, a native of the state of Maine who represented a group of Eastern capitalists from the United States. Hydro electric power could not then be carried any great distance over transmission lines; therefore, industries had to be built near the source of power. One of the greatest entrepreneurs of his time, F.H. Clergue, was quick to visualize the future of both the Canadian Sault and the American Sault as industrial centres as a result of the availability of cheap hydro power. He promoted a number of enterprises on the Ontario side of the river, including a steel mill. He built a railway, developed iron mines and had his own fleet of steamships to carry his iron ore; but like many pioneer promoters of the time, he was destined to see his vast enterprises collapse. The dream of making the two centres into industrial communities was taken up by others though F.H. Clergue can be truly called the father of the industrial Sault Ste. Marie.

Construction on a 100-ton groundwood mill at the Sault began in 1895. It began to operate a year later. Bertrand J. Clergue, a brother of the founder, was manager. When the company went into liquidation in 1903, the mill was taken over by a new firm called the Sault Ste. Marie Pulp and Paper Company Ltd. In 1911, George H. Mead of Dayton, Ohio purchased the original groundwood mill and a sulphite mill, which had been added to it. The name was changed to the Lake Superior Paper Company Ltd. In 1913 it was amalgamated with the Spanish River Pulp and Paper Mill Ltd. In 1928, it was in turn merged with the Abitibi Power and Paper Company Ltd. By 1960, this plant now had four newsprint machines with a daily capacity of 305 tons and a five cylinder board and wrapper machine, turning out 100 tons per day of unbleached sulphite.

The third mill to be built in Northern Ontario was that of the Spanish River Pulp and Paper Company at Espanola. The directors of this enterprise were W.J. Shepherd of Waubaushene and Thomas Shepherd, J.J. McNeil and James Tudhope, of Orillia, Ontario. The construction contract was awarded to the firm of Munroe and Piggott and erection of the power dam began in 1899. The following year the company started to build a pulp mill which was in operation by 1905. In 1910, the company was incorporated under the name of the Spanish River Pulp and Paper Mills Ltd. In 1911, a paper mill was added, which began production in 1912. In 1928 the company was absorbed by the Abitibi Power and Paper Company and the plant was closed down in 1930 during the Great Depression.

In January, 1943, the Kalamazoo Vegetable Parchment Company of Kalamazoo, Michigan, acquired the Espanola plant from Abitibi with their timber rights and water power, and began on an extensive program of construction. A sulphite mill was erected and was in operation by the summer of 1946. Meanwhile, the existing ground wood mill had been reopened, with a daily output of 30 tons per day. Production of the bleach plant was accelerated to the daily rated capacity of 200 tons of bleached kraft. By 1960, two paper machines were scheduled to be in operation shortly, with two more to be made available when needed. This company was incorporated in Canada in 1946 as the KVP Company Ltd. to take over the assets of the Abitibi Corporation at Espanola. They had cutting rights over an area of approximately 5,500 square miles of timber lands. This took in all the various species of coniferous and deciduous woods. The company employed some 1,800 workers in their mills at Espanola and their woods operation. A number of dwellings were added to the townsite, as well as other needed public buildings assuring the town of Espanola of a bright future.

The parent company was founded at Kalamazoo, Michigan, in 1909, by the late Jacob Kindleberger, generally known as "Uncle-Jake." Its object was to convert waterleaf into genuine vegetable parchment. In 1910, a waxed paper division was added. This was followed by further constantly expanding operations, which finally necessitated the construction of three additional paper mills. The first was erected in 1918, the second in 1923, and the third in 1928, with a total daily capacity of 300 tons. Jacob Kindleberger, who died in January, 1947, was a man of order. Rather rigid in his habits, he insisted on discipline from his subordinates. The two towns of Kalamazoo and Parchment, which adjoins it, are well built, attractive in appearance and well administered. The company is mainly engaged in manufacturing food protection papers.

It has subsidiary companies in Pennsylvania and Texas, as well as Sturgis, Michigan. In Canada, it also acquired the interests of the Appleford Paper Products Ltd. of Hamilton, and Aridor Ltd. of the same city.

The next development in the provincial pulp and paper industry was at the extreme northwest of Ontario. The Fort Frances Pulp and Paper Company Ltd. had been organized as a subsidiary of the Backus Brooks Company which held the power rights on the waterfalls that form the headwaters of Rainy River, an International waterway dividing Northern Minnesota from Ontario. The company was committed, by agreement with the Ontario Government, to build a mill on the Canadian side. It began to operate in 1914. The name of the Company was subsequently changed to that of Backus Brooks Company, and eventually to the Ontario and Minnesota Power and Paper Company. It was part of the system which operated mills at International Falls on the American side and in Fort Frances and Kenora, Ontario.

In 1912, the company obtained 1,800 square miles of timber lands from the Ontario government, on the understanding that they would erect a mill near or about the town of Kenora. In 1920, they obtained from the Drury government a further timber concession of 3,000 square miles. Their plans for the construction of a paper mill at Kenora were prepared and, with financial arrangements completed, they began the construction of a 350-ton newsprint mill in 1920 which was fully completed and in operation two years later.

Concurrent with the establishment of a paper mill by the Backus Brooks Company on the Canadian side of the river at Fort Frances in 1914, a 30-ton pulp mill was erected at Dryden, Ontario. This enterprise, the Dryden Timber and Power Company, was promoted by Lester S. David, a lumber manufacturer from Seattle, Wash., and E.W. Bonfield of Wisconsin Rapids. It began to operate as a small sawmill in 1910 and was shortly afterward put into liquidation. It was re-organized by the two main stockholders, David and Bonfield, and refinanced by Lazard Freres and Company Investment Bankers with offices in Paris, London, and Montreal. It was then named the Dryden Pulp Company with Bonfield as manager. Originally its production was confined to pulp, subsequently it began to manufacture wrapping, building paper, paper bags and kraft products. Eventually it was renamed the Dryden Pulp and Paper Company Ltd., and then the Dryden Paper Company. Their former general managers appeared in the following order: E.W. Bonfield, G.B. Beveridge, William Bullard, J.S. Wilson, and Lloyd Bruce. By

1960, it had a yearly output of 50,000 tons of paper and 92,000 tons of pulp, bleached and unbleached. This mill has been an asset to the entire district of Dryden, furnishing settlers with a ready market for their wood products and the produce of their farms, in addition to giving employment in 1960 to 700 persons in and about the mill, and an additional 500 workers in their woods operations. Under the skillful leadership of E. Lorne Goodall as president, and by his successor T.S. (Tommy) Jones, general manager and more recently president, this mill has been completely modernized and its capacity substantially increased making it one of the best equipped and operated kraft mills in Canada.

The Abitibi Pulp and Paper Company followed with the construction of a mill at Iroquois Falls. This enterprise was promoted through the vision of F.H. Anson and largely financed on its future earnings. Frank Cochrane, while Minister of Lands, Forest and Mines, had made several attempts to interest manufacturers and financiers in taking up pulpwood concessions and waterpower rights in Northwestern Ontario. F.H. Anson became interested, and with Shirley Ogilvie of Montreal, began the erection of the mill in 1912. In the late 1920s there was an interesting editorial in the *Globe* entitled "Romance in Financing" describing the various steps taken to construct the original Iroquois Falls enterprise, how it was able to meet all its obligations, and also pay regular dividends on both the preferred and common stocks. In 1914 its charter had been surrendered to the Abitibi Power and Paper Company, who took over the assets of the original firm. By 1960, this large mill at Iroquois Falls enjoyed a continent-wide reputation for its high-quality products. It operated seven newsprint machines and one wrapper machine, with a daily capacity of 625 tons of newsprint and 35 tons of wrapper. It employed thousands of men in the combined mill and woods operation. Its townsite at Iroquois Falls was one of the best arranged community centres of its size in Northwestern Ontario. Its power plant at Iroquois Falls, on the Abitibi River, developed 28,000 hydraulic horse power; that at Island Falls, 48,000 electric horse power. It also operated a power plant at Twin Falls, with a capacity of 30,000 electric horse power.

One of the first cuts of pulpwood for the Abitibi Power and Paper Company was taken out during the winter of 1904-15 by R.O. Sweezey. His operations were on the Low Bush River, north of Lake Abitibi, and consisted of 20,000 cords all delivered in the spring of 1915 in boom at the lake, at a cost of $3 per cord to the company. Low as that price was, the contractor netted a small profit.

Another development promoted at Smooth Rock Falls was under the name of the Mattagami Pulp Company Ltd. It originated in 1916 with Duncan Chisholm who for two years previously had been buying veterans' settlement claims in the Matagami River Valley, south of the transcontinental railroad. These claims, which he seems to have purchased at a very low price, represented about 100 square miles. He then secured a concession of some 800 square miles of lease hold timber limits along the Matagami River, which he added to his own in fee simple lands. With Sherwood Aldrich, formerly a mining man from Arizona, he erected a sulphite pulp mill at Smooth Rock Falls, after which the town was named. A.G. McIntyre became the general manager of this mill and remained in charge for about a year and a half. Shortly afterwards this venture got into financial difficulties. The Royal Securities Corporation Ltd. had bought out the first mortgage bonds in 1916. R.O. Sweezey was engineer for this banking house, and had a great deal to do with the supervising of the operations for a time, on behalf of the bondholders. He took out a contract for their pulpwood, in partnership with J. McGrosart, a former woods manager of Price Bros. & Company Ltd. Their men were brought up from the Saguenay River, and in time became settlers in the district. According to Sweezey, wood was delivered at the mill at $4 per cord. In the late 1920s this firm was taken over by the Abitibi Corporation. At present (1961), the Smooth Rock Falls mill is devoted exclusively to the production of high-grade, bleached sulphite pulp for the fine paper specialty market. It has one up-to-date drying machine with a daily capacity of 210 tons. Its own power plant at Smooth Rock Falls develops 9,500 hydro electric horse power.

The Port Arthur Pulp and Paper Company Ltd., now Provincial Paper Ltd., was next to be organized, with construction commencing early in 1917. The mill was in operation a year later. It was enlarged in the early 1920s.

It was around this time that another promotion was floated in Port Arthur, namely the Kaministiquia Pulp and Paper Company Ltd. They built a small pulp mill on the right bank of the Current River, south of the Canadian National Railway tracks. The president of this organization was U.M. Waite, and G. Regan was secretary-treasurer. C.D. Howe of Port Arthur was made a director and his firm took on the job of constructing the mill. Pretty well completed, it went into liquidation in January, 1921. Its superintendent was C.L. Parsons and Erle Smith, later of the Department of Highways, was engineer in charge of construction. This plant was finally sold by agreement dated July 25, 1921, to George P. Berkey of Wisconsin Rapids, acting on behalf of the Consolidated Water, Power and Paper Company.

Concurrent with the construction of the Kaministiquia Pulp and Paper Company Ltd., another enterprise was floated on the Mission River at Fort William by T.R.H. Murphy, agent for the Mead Corporation of Dayton, Ohio, headed by George H. Mead. It was a groundwood mill of 50 tons. Construction began in the fall of 1920 and it was completed and in production by the fall of 1921. T.R.H. Murphy had been in charge of the construction of the Port Arthur Pulp and Paper Mill and was then on his own as a consulting engineer with an office in New York (where, as of 1961, he was still in business). The price of pulp eventually dropped to the extent that it was unprofitable to continue operations. The price of newsprint, however, commanded substantial profits; so it was decided by the board of directors to convert the mill to the production of newsprint. Norman M. Paterson was a director of the enterprise along with the late W.A. Black, director of the Kam Power Company Ltd., now owned by the Hydro Electric Power Commission (Ontario Hydro). Together with the other directors, they arranged with Sir Herbert Holt to reorganize the company, which had its name changed from Fort William Pulp to Fort William Paper Company Ltd. Additional capital was subscribed for the installation of a paper machine. It operated steadily until 1928 when it was sold at substantial profit to the Abitibi Power and Paper Company. Like all mill promoters of the time, T.R.H. Murphy had obtained large timber areas from Crown lands before undertaking to go ahead with his enterprise. His timber limit formed a part of the timber areas held by the Abitibi Power and Paper Company. As of 1960, the plant had two newsprint machines, with a production of 195 tons per day utilizing power from the Kaministiquia Power Company.

As already noted, the 1920s saw tremendous speculation all over the continent. Large organizations were acquiring footholds in the virgin forests of Northwestern Ontario, from which they would be able to assure abundant and continuous supplies of pulpwood, and were preparing plans for the establishment of mills of their own near these forest stands. This led a number of local timbermen and speculators to see what they could do for themselves in acquiring timber rights.

A partnership was formed between Walter H. Russell and Colonel

J.A. Little and a charter obtained in 1922, under the name of the Ojibway Timber Lands Company Ltd. They succeeded in having the Township of Hele near Nipigon, one of the finest timber limits in the district, set aside for themselves. General D.M. Hogarth then joined the syndicate, but disposed of his third interest shortly after to Provincial Paper Company. Subsequently, the remaining two-thirds interest in this timber limit, held by Russell and Little, was sold at a substantial profit to Provincial Paper. The same group then organized the Nipigon Fibre and Paper Mills Ltd., and erected a small pulp mill on the shore of the bay near the village of Nipigon, with a capacity of 80 tons of mechanical pulp a day. Walter Russell had contracted to supply this plant with pulpwood. From all accounts, it was not a profitable undertaking, but it kept above water until 1926, when the company was taken over by the Nipigon Corporation Ltd. The Nipigon Corporation, headed by N.A. Timmins, president, and J.I. Rankin, secretary-treasurer, had entered into an agreement with the Department of Lands and Forests on January 13, 1926, to begin immediately on the construction of a mill with capacity of 350 tons of mechanical pulp per day. It also agreed to build a sulphite mill capable of turning out 125 tons of sulphite per day, and a paper mill with a production of at least 400 tons of newsprint each 24 hours. In view of this undertaking, the company was granted vast area of choice timber land, and purchased the small plant of the Nipigon Fibre and Paper Company Ltd. for $1.5 million. It was one of the juiciest timber transactions of that period. For some reason, Walter Russell was left out of it. The beneficiaries were Colonel J.A. Little and Donald Hogarth. It was from this profitable transaction that Hogarth got his financial start.

The Timmins interests immediately sold out, most likely at a very high profit, to the International Power and Paper Company Ltd., who likewise made next to no attempt to carry out the undertaking agreed upon between the Timmins interests and the Department of Lands and Forests. Some years afterwards, the International Power and Paper Company was in financial difficulties, but managed to operate the small mill with wood from their fine timber stands. Then R.O. Sweeney organized his Lake Sulphite Company at Red Rock in 1937. He bought out all interests of the Nipigon Corporation, which later formed a part of the holdings and plant of the St. Lawrence Corporation Ltd. It would seem evident that the first two transactions were purely speculative. The first syndicate, representing the Nipigon Fibre and Paper Mills, and the second one the Nipigon Corporation, had both greatly profited. The International Power and Paper Company were thus the owners of large timber areas, but never endeavoured to carry out the obligations under-taken when this vast timber limit was obtained from the Crown. The International Power and Paper Company were at that time riding high. They were dominating the paper market of the United States. By a stroke of the pen, they could dictate the price for all competing mills. Their empire was being pyramided year after year, if not from month to month, in the financial markets of New York, Boston and Montreal. According to a 1928 prospectus, which was circulated widely in Canada and the United States, they had timber rights in New Brunswick, all across Northern Quebec and as far West as the Nipigon district. They could boast of having under lease timber limits representing a total area equivalent of three states of the Union. In this blurry timber transaction from Crown lands to private ownership in the Nipigon district, no one had benefited to any extent except the speculators.

When the Consolidated Water Power & Paper Company of Wisconsin Rapids took over the Kaministiquia Pulp and Paper Company mill at Current River, they took out a charter under the name of the Thunder Bay Pulp and Paper Company Ltd., and began to operate their small mill. In 1926, they decided to enter the field as manufacturers of newsprint in Port Arthur, and having obtained sufficient timber limits, began the erection of a modern plant at a cost of $5,900,000. This went into operation in 1927. It was sold a year later to the Abitibi Power and Paper Company and operated as the Thunder Bay Paper Company Ltd. It became a thoroughly modern plant, with two newsprint machines and, by 1960, a daily production of 295 tons.

The president of the Consolidated Water Power and Paper Company was George W. Mead. There were two Meads who played leading parts in the development of pulp and paper products in Northwestern Ontario. One was George H. Mead, president of the Mead Corporation of Dayton, Ohio, who, as has been noted, reorganized the Sault Ste. Marie Pulp and Paper Company Ltd. in 1911, and was the prime mover in the organization of the Fort William Pulp and Paper Company in 1920. The Mead Corporation are large brokers in pulp and paper products. Since the two names occur so often together in North and Northwestern Ontario, the timbermen, to avoid confusion in referring to them, used to call George W. Mead, president of the Consolidated Paper Company, George Wisconsin Mead, or simply Wisconsin Mead. They were thus able to tell at a glance which one of the two Meads was involved in this or that transaction. Both of them have undoubtedly made a distinct contribution towards the development of our forest industries.

As early as 1921, E.W. Backus, head of the Backus-Brooks enterprises, had already decided to become involved in the largest of all the mills at the Canadian Lakehead, the Great Lakes Paper Company Ltd. For this purpose he organized subsidiary companies and acquired timber rights to vast portions of Northwestern Ontario.

It might be well first to review the spectacular career of this unusual man who, from a humble beginning, built a mighty industrial pulp and paper empire and then witnessed its collapse. He was man of unusual vision, extraordinary energy and undaunted determination. He had unbounded faith in himself and an infectious enthusiasm for any project that he undertook. Unsparing of his own efforts, he demanded unstinted effort from all who were associated with him. He could never comprehend why they were not able to match him in both energy and action. He did not waste much time in catering to public opinion. It could almost be said of him that he embodied the famous phrase attributed to Vanderbilt half a century previously, "The public be damned." Backus had had considerable experience in dealing with legislators in his own state of Minnesota before coming into Ontario. He was able to make good use of politicians, irrespective of their party labels; he believed that every politician had his price and he was reported to be generous at election time. There was a great deal of the showman in him. He publicized his various enterprises in the most unique ways. On one occasion he gave a clam bake party on one of the islands of Rainy Lake. For this picnic he brought the clams, and even the sea weeds, which were needed to give the clams their proper flavour, from the coast of Maine. He brought in a brass band from the United States, and invited Indian chiefs. For two days it was a real pow-wow. He even invited financial editors, editorial writers and paper manufacturers from the United States, and they painted the district red. Harry McMahon, formerly president of the Port Arthur Beverage Company, a Bostonian who knew good seafood and how to prepare it, was in charge of the menu on that historic occasion.

Backus was a man of culture, and a lover of the arts. His home life was simple. He was no "social climber." His obsession was planning more and more industrial developments and acquiring greater and greater timber limits. He could make quick decisions, usually accompanied by plucking one of his eyebrows, a mania which indicated a highly nervous temperament. E.W. Backus could be acknowledged as one of the last, aggressive, free-booting enterprisers of the early twentieth century.

Dan McLeod, an all-round sawmill man with considerable knowledge of timber operations, was general manager of all the Backus owned mills in Northwestern Ontario. This included the plants at Fort Frances and Kenora and the Great Lakes mill at Fort William, as well as sawmills at Keewatin and Hudson. Dan McLeod began his career the hard way. In summer months, he worked as a sawyer for the Mathers Lumber Company, and in the winter months as a log scaler. When the Mathers Company was acquired by Backus-Brooks, McLeod was appointed manager. Dan McLeod was neither brilliant nor spectacular. He was a plain man, unassuming and methodical; a plodder. He never knew what it was to take a vacation. Rather rigid in his habits, he was always referred to by his intimate friends as "Tea-kettle Dan." He was more fond of tea than of liquor. He was a good family man and a good citizen. A typical Scotch Presbyterian, with a host of friends, he commanded respect. He died in 1946 in his eightieth year.

The superintendent of woods operations for Backus-Brooks was Captain George McPherson. A veteran of the First World War, often referred to as Colonel McPherson, he was associated with Dan McLeod practically all his life. McPherson was the most colourful figure of his day in and about Kenora and Fort Frances. Of striking appearance and distinguished bearing, he looked equally well dressed in cork shoes and a mackinaw, or with a Stetson and the latest broadcloth clothes made to order by the best of London tailors. His diamond ring and his well trimmed moustache and Vandyke beard, completed his personal makeup. To women, he was a heart-throb, but he never married. He was as much at home with the bushmen as he was with the notables. He is reported to have made a great impression on the Prince of Wales when they met at the reception in the Royal Alexandra Hotel in Winnipeg in 1919. Both George McPherson and Dan McLeod had been deeply devoted to E.W. Backus who, in turn, had implicit confidence in their judgment. It is quite likely that it was on the recommendation of E.W. Backus that both of these loyal executives were provided with generous pensions by the liquidators of the company.

E.W. Backus' promotion of his many enterprises reads like a romance. In 1884, he had already accumulated some capital and was able to purchase the lumber firm of Lee & McCulloch in Minnesota for

$6,000. The name was changed to E.W. Backus & Company. In 1892, without the investment of added capital, the total assets of his firm represented approximately $600,000. He reorganized a firm under the name of E.W. Backus Lumber Company, and took over the assets of the former partnership. He then began to add affiliated syndicates and incorporated the Minnesota Logging Company, acquiring extensive pine timber limits containing approximately two billion feet of lumber. He then purchased the Brainard and Northern Minnesota Railway Company (later the Minnesota and International Railway Company) which he extended to his timber limits. Eventually this railway was constructed to Bemidji, Minnesota. In 1899, William F. Brooks became a member of the firm and a company was organized under the name of the Backus-Brooks Company.

Early in 1900, they acquired a controlling interest in the Koochiching Realty Company, which held large holdings of timber lands in Northern Minnesota, adjoining International Falls, across from Fort Frances. Backus then opened up negotiations with the Ontario government and acquired the water power site and land on the Canadian side of the river, with the view to establishing a mill. He purchased additional lands on the American side for a mill site and railway yards to further the development of water power and the construction of mills on both sides of the river. In 1902, Backus organized the Backus-Brooks Company, under the laws of Maine. This firm took over the Minnesota assets held by the old firm. Their financial assets had grown to $3,000,000. In 1903, he organized the Namakan Lumber Company in Minnesota, which company acquired still vaster timber holdings in Northern Minnesota. In 1905, the International Lumber Company, a wholly subsidiary organization, was founded with a fully paid up capital of $600,000, later increased to $4,000,000. The company also organized the First National Bank of International Falls and the International Telephone Company. Then they organized the Rainy River Improvement Company and the Minnesota Power Company Ltd. The capital stocks of these companies were fully paid for and were owned by the Backus-Brooks Company. Construction began in 1905 on the water power plant and paper mills at International Falls.

Never idle a moment, Backus in 1906 organized the Keewatin Lumber Company, which bought out the interest of the Mather Lumber Company Ltd. at Keewatin and acquired their water power rights on the Winnipeg River, known as the Norman Dam. The Big Fork and International Falls Railway Company was then organized to be taken over by the Northern Pacific Railway, and as we have already seen, Backus began the construction of a paper mill at Fort Frances in 1911.

In 1914, with the objective of utilizing all by-products and wastes from both his sawmills and paper mills, Backus founded the International Insulation Company, with a paid up capital of $50,000. He bought out the patented process from the inventor and immediately began production of pulp board, which became known as insulite. This organization was a subsidiary of the Ontario and Minnesota Power and Paper Company, which in time became known as the Insulite Company. By 1919, without additional capital, the original investment of $50,000 had expanded to $5,000,000 with an annual profit of $1,000,000.

By 1920, as we have already seen, Backus began construction of a paper mill at Kenora. As early as 1921, Backus had already decided to take an interest in the largest of all the mills at the Canadian Lakehead. For this purpose he organized the Transcontinental Development Company Ltd. as a subsidiary company. He had purchased large areas of timber lands on the Black Sturgeon and Pic Rivers from the J.J. Carrick syndicate, and, through Trancontinental Development, took over the Nagagami timber limits estimated to contain 5,000,000 cords of pulpwood, from J.H. Black, in addition to some 600 square miles of private timber land in fee simple. With these resources in hand, Backus was now prepared to finance a major undertaking which was named the Great Lakes Paper Company Ltd. In 1923, he erected a pulp mill on the present site of [Avenor's] Thunder Bay mill. In 1928, he added a large paper mill to it, which began operation the following year. This mill was constructed under the supervision of J.T. McLennan, his construction engineer. J.H. Black, from whom he had purchased his Nagagami timber limit, was the first superintendent of operations. This was the largest of all his mills, and one of the most modern plants on the continent. By 1960, it was producing 1,000 tons of newsprint per day and 20,000 tons of pulp products annually with 1,300 persons employed in the mill and 800 in their woods operations. The annual payroll at the time was $6,750,000; that of the various forest operations amounted to $2,200,000. It had become the largest newsprint mill, as well as the largest sulphite mill, in Northwestern Ontario.

By the time the mill was completed in the 1920s, however, the boom was over and, by 1932, the Backus-Brooks industrial empire, like many others, had collapsed, a victim of the Great Depression. Like most enterprises of that extravagant and speculative period that preceded the 1930s, it was not to be reorganized without considerable loss to the stock and bond holders. The 1920s witnessed a great battle among promoters and speculators for the control of forest lands in Northwestern Ontario. By 1930, the Minister of Crown Lands, the Hon. William Finlayson, who had succeeded James Lyons in 1927, stated in his report on the Lake Superior region that of the total forest area of 5,956,000 acres, 222,000 acres were held under timber license and 4,821,000 acres were included in pulpwood concessions. Likewise, in the Nipigon extension, which contained 1,753,700 acres, 1, 522,560 acres were under pulpwood concessions and 10,240 acres under timber licenses. Practically the entire forest resources of the Lake Superior region and the Nipigon extension were held, at this time, either by permit or by concession by a few companies or speculators. Yet, at that time, there were only four pulp and paper mills at the Canadian Lakehead, namely Fort William Paper, Great Lakes Paper, Provincial Paper and Thunder Bay Paper. The small plant of the Nipigon Corporation was manufacturing pulp at Nipigon. It had evidently been a prosperous period for the Timber Wolves, which was only to be corrected by the depression of the 1930s.

If there is one industry that the Canadian people should be most efficient in managing and highly successful in operating, it is the timber industry. Our pulpwood reserves are fortunately located along waterways that are crossed by three railways and the vast areas of timber stands are along most of the rivers flowing into the Great Lakes and St. Lawrence drainage system. Our black spruce pulpwood ranks among the best in the world. Our woodsmen have had generations of training in woodcraft of all types. In addition to developing our own lumber industries, the Canadian bushmen contributed largely to the development of the Northern State bordering on Canada; Canadian woodsmen were always in demand at the time that these forests were exploited. Our winters lend themselves to economical operations. Everything is in our favour. Yet, by 1927, there was evidence that the pulp and paper industry was heading for a slump. The capacity of Canadian newsprint mills was being doubled, while the consumption of newsprint in the United States had increased by only 20%. The market was over saturated. It was at this time that an editorial appeared in the *Montreal Star*, laying the blame for the industry's difficulties at the door of Premiers L.A. Taschereau of Quebec and G. Howard Ferguson of Ontario. There is no doubt that both Quebec and Ontario had granted far too many timber concessions to promoters and speculators. It may have been just a coincidence, but these timber concessions associated with paper mill construction were usually handed out at election times when political parties were badly in need of campaign funds. It will be the purpose of the next chapter to examine some of the causes that contributed to the slump in what should have been Canada's soundest and most prosperous industry.

Chapter Seven

Depression and Recovery

P eriods of intensive speculation have usually been followed by panics, crises, and depressions with attendant terms of disruption and
unemployment. Speculations have never enriched communities,
only a few speculators. Boom periods and high prices are usually associated with loose management, short-sightedness, and extravagance in
the operation of any enterprise. Since the beginning of the 19th century
until the Great Depression, there were no fewer than fifteen successive
crises. These did not happen, however, at regular intervals; in short,
they did not recur with mechanical regularity. There were periods of
prosperity ranging from three to sixteen years. The average length of
each period was approximately eight years.

A glance at the records will show that panics have always followed
the spectacular collapse of what was believed to be a sound banking
house, a railway company, or an industrial organization. Some examples include the banking house of Jay Cooke and Company in 1873; the
Philadelphia Reading Railway and the National Cordage Company in
February, 1893; the Knickerbocker Trust Company of New York City
in 1907; and, in Canada, the Peter Lyall Construction Company in September 1929. None of these were the cause of a panic, but simply
indicators of unsound financial conditions existing in the North American economy at the time. Fortunately, panics are followed by corrective
periods which again bring people to their senses. Economists affirm
that there is no substitute for careful planning application and thrift; as
the late Theodore Roosevelt once wrote, "there are no substitutes for
sustained thrift, industry, application and intelligence."

In the 1920s, speculators, promoters and investment bankers had
violated almost every tenet of orderly business common sense. This
applied to all the North American continent and to every phase of industrial and financial activity. In Western Canada, for example, uncon-
trolled wheat prices after the First World War made land values sky-
rocket; the crash that inevitably followed in the early 1920s was particularly dramatic. It was said by many observers in Western Canada
that farmers acquired substance and ease in hard times which were dissipated in periods of boom. Human nature is no different, whether a

man is a farmer, an industrialist, an investment banker, a speculator or a promoter. Lack of foresight and extravagance leads, generally speaking, to panics and crises, followed by depressions such as that witnessed during the 1930s.

Paper mill promoters and timber speculators were not spared that common error. Here is an illustration: a manufacturer has a well established plant of which he and a few friends are the owners. Under his management and that of his associates, his mill has operated successfully over a period of years. He has met all his obligations when he had occasion to borrow from banks, has paid dividends to the stockholders, and has the confidence and goodwill of his clients wherever his products are sold. When he is then approached by a promoter seeking to buy his plant, the owner replies that he is prepared to put a value of, let us say, $5,000,000 on his total assets, including the goodwill of his company. An option is secured from the owners. Other companies are brought in by similar methods. Then the promoter goes to investment bankers, and after a number of conferences, organizes a new corporation. Having added fictitious values, a charter is obtained, attractive booklets and circulars are prepared and sent out to every nook and corner of the country. Fine advertisements are displayed in the press, and the public, not realizing that they are contributing to a scheme which has discounted the future from 25 to 30 years or more, subscribe to the bond issues, and purchase preferred, common, voting and non-voting stocks. Then this large combine is floated and attempts against great odds to operate profitably.

This would seem to be an exaggerated story. Unfortunately, it is true. It happened over and over again during the 1920s. It happened even before that. It will happen again unless governments take means to regulate the greed and cupidity of promoters, brokers and investment bankers. There is a vast difference between control and regulation. Cities regulate traffic on their streets and governments regulate traffic on their highways. Red and green lights and stop signs are installed to prevent automobile and truck drivers from running wild. Yet, accidents do happen. Stop signs and red lights will have to be installed to prevent overly optimistic and irresponsible promoters, brokers and investment bankers from organizing and re-organizing enterprises with resultant ill-effects on the investors and the workers alike. In Northwestern Ontario, there are three prime examples of companies that over-estimated the markets and mortgaged their futures: the Abitibi Power and Paper

Company, and the Backus-Brooks Company, and the Lake Sulphite Company. Extravagantly promoted schemes were common all over the continent and contributed in a great measure to bring about the Great Depression. The pulp and paper industry became, therefore, no longer an industrial development, but a financial operation controlled largely by investment bankers who were ill fitted to run such an important technical industry. In order to create profits on exaggerated capitalization, they had recourse to two methods. First, they brought in efficiency experts and outstanding engineers in an attempt to intensify production, and thus over saturated the market. Second, they imposed rigid price controls and divided the trade, the very antithesis of the capitalistic system, the principles of which are individualism, private initiative and freedom of trade.

Capitalism carries within itself the very germ that creates its ills. If the disease is not checked, it may eventually lead to destruction. In diagnosing the malady, one thing appears clearly evident. The Great Depression was brought on and was unduly prolonged because capitalists had violated two essential principles of free enterprise: they created too much fictitious capital by unwisely discounting the future. And, equally seriously, they attempted by various devices to prevent, or at least to limit competition.

The combines, trusts and cartels that were formed to control prices and divide up markets, eventually collapsed. This, in turn, led governments to intervene and set up quotas. In both Ontario and Quebec the system of prorating was introduced. In time, even the League of Nations made an attempt to effect some measure of relief. Thus, as it was pointed out by the eminent British Economist, Graham Hutton, "we were witnessing the paradoxical spectacle of a group of nations trying to restore freer trade throughout the world by government control."

When Franklin Roosevelt accepted the Democratic Party's nomination at Chicago July 12, 1932, he referred to this wild speculation:

> What was the result? Enormous corporate surplus piled up, the most stupendous in history. Where, under the spell of delirious speculation, did those surpluses go?... Why, they went chiefly in two directions: first, into new and unnecessary plants which now stand stark and idle; and second, into the call money market of Wall Street, either directly by the corporations, or indirectly through the banks. Those are the facts. Why blink at them?

And again, in his inaugural address of March 4, 1933, he stated: "The money changers have fled from their high seats in the temple of our civilization. We may now restore that temple to the ancient truths. The measure of the restoration lies in the extent to which we apply social values more noble than mere monetary profit." In a letter dated Jan. 9, 1934, to his old, loyal employees, E.W. Backus, president of Backus Brooks and Minnesota and Ontario Paper Company Ltd., recounted what he had accomplished during his lifetime. He stated:

> When the crash of 1929 swept the country, the newsprint industry was feeling keenly the effects of overproduction. However, there was neither hesitation nor fear on my part. The thought of any default on outstanding obligations never entered my mind — a fact perhaps best attested in that approximately $3,000,000 was expended during 1930 in expanding our properties in the United States, Canada and Europe. Looking back now, I can see where I made one mistake. I placed my reliance in investment bankers. At that time I had no more conception of their inherent untrust-worthiness than that possessed by the public generally.

These statements, one from the chief executive of a great nation and the other from an outstanding industrialist, clearly support the contention already made that the development of our primary industries was no longer regarded as purely industrial investments, but as financial operations.

This investment climate in the 1920s was not unprecedented. Some decades earlier, the construction of the Grand Trunk Pacific Railway was regarded by investment bankers not so much as a railway construction, but as a financial operation, a fact they now concede. When the bill authorizing funding for the railway was presented to the House of Commons by Sir Wilfrid Laurier, then Prime Minister of Canada, his Minister of Railways, Andrew G. Blair, resigned largely on that account. Defending his decision, Blair hurled this famous aphorism at his former chief: "Cox can't wait." He was referring to George A. Cox, a leading financier in Canada, president of a life insurance company and a director of investment banking institutions, who had a great deal to gain in the promotion and construction of the Grand Trunk Pacific, which, for a period of 50 years, was to be a problem.

It is doubtful if politics has been played harder in any other part of Ontario than at the Canadian Lakehead. James Conmee reigned supreme as an M.P.P. until 1904, and then as an M.P. until 1911. During that time, whether at Toronto or at Ottawa, he exercised an influence more pronounced than that of some cabinet ministers. "Fighting Jim," as he was known, was a power to be reckoned with in his day. It is questionable, however, if in his time of glory, he ever exercised as much power as Conservative Don Hogarth. Hogarth's influence continued even after the defeat of the provincial Conservative government in 1934. Recognizing the realities of political power in Northwestern Ontario, Hogarth quickly adapted himself to the change of government and became firmly attached to Peter Heenan, Hepburn's new Minister of Lands and Forests. They became fast friends; it was even gossiped that Hogarth had contributed generously to Heenan's campaign funds.

As early as 1928, Howard Ferguson had indicated his desire to withdraw from political leadership, although he held the confidence of both his party and the electorate. He had accomplished most of what he had set out to do. As a minister in the Hearst Government, it was his administration of lands and forests that had been so closely scrutinized by the Royal Timber Commission. There was a belief across the country that he was aiming at the leadership of the federal Conservative party in 1927, although he himself had never indicated this. To this suggestion, so one story goes, a staunch party member from Toronto had said: "Ferguson is an able leader, but he still smells too much of the odour of spruce trees to be named the leader of our party." At any rate, during his tenure of premiership, when the Honourable James Lyons was Minister of Lands and Forests and had a great deal to do with new highway development in Northwestern Ontario, there was considerable criticism expressed in and out of the House about the propriety of a cabinet minister who, on account of his large business connections, was bound to benefit indirectly from the granting of both the highway contracts and timber permits and pulpwood concessions. From his seat in the legislature, Howard Ferguson made the suggestion that no cabinet minister with such a setup should remain in the cabinet. James Lyons, who was operating a large supply yard and hardware store at Sault Ste. Marie and was engaged in both wholesale and retail business, resigned. But to his credit he remained loyal to his party. He was succeeded in 1927 by William Finlayson, member for Simcoe, who gave the department a fairly honest, if not brilliant, administration.

In the early 1930s, Howard Ferguson resigned as leader of his party, and was shortly afterwards appointed high commissioner to London, a post which he filled with skill, tact and dignity. He was succeeded as Premier by George Henry, another straight-laced man with limited ability and absolutely no vision. "Honest George," as he was called, was not a formidable leader. When his party went down to defeat, many of his friends blamed him. It was said of him that he was stupidly honest. Some uncharitable followers even stated that if George Henry had to cross a field in which there were a number of cow pats which he could have avoided, he was sure to put both feet in every one. It was Don Hogarth's task to put Henry over as typifying "Old Man Ontario," but despite all of Hogarth's gifts as an organizer, the party foundered in 1934. The Liberals were elected by a large majority with Mitch Hepburn at their head.

When the Liberals assumed power, conditions were undoubtedly bad. Sir Wilfrid Laurier is reported to have once stated that after eight years of power a government begins to show signs of decay and corruption. The Conservative administration had held power from 1923 to 1934, and abuses and patronage had, to a great extent, crept into the administration. Such is the inevitable outcome of our party system. Mitch Hepburn had a splendid opportunity to prove himself as a real leader of a reform party, and for a while he shot both his guns in every direction, getting rid of a lot of hangers-on. It was not long, however, before his party also began to gather barnacles, a fact that became very evident when he appealed to the people for a second time in the fall of 1937.

Hepburn's administration of Lands and Forests was, at first, a continuation of the policies of the previous government. Over time, the long-held policy of prohibiting the export of pulpwood to U.S. markets was gradually eroded. Prior to 1932, very little pulpwood had been exported from Crown lands. Until 1931, exports were confined mainly to poplar, which was shipped by the authority of the minister under a 1919 Act. In 1920, this Act was amended to grant the lieutenant-governor in council the authority to suspend the manufacturing condition and to permit the export of pulpwood of any species cut on Crown land. Until 1931, this provision was to be invoked only in exceptional cases, specifically to export spruce and balsam pulpwood. In February of that year, however, an Order-in-Council was passed making pulpwood exportable from Crown lands, provided substantial clearances for a like quantity of similar wood cut on patented or settlers' lands, was supplied to paper mills in Ontario. Between 1931 and 1934, additional Orders-in-Council were passed, granting individual contractors the right to export pulpwood from Crown lands, on condition that the usual substitution of clearances was furnished to the department.

By an Order-in-Council dated May 16, 1933, the manufacturing conditions for that year were entirely suspended for peeled pulpwood cut on Crown lands in the district of Thunder Bay, provided substantial clearances for a like quantity of exportable wood was shipped to Ontario mills. This was further extended to 1934 and 1935. On March 23, 1935, a General Order-in-Council was passed, permitting the export of pulpwood of any species cut on Crown lands in Ontario, provided contracts were supplied, and the Minister of Lands and Forests was satisfied that the pulpwood so exported would not be used for the manufacture of newsprint and would not compete with established Ontario industry. Similar Orders-in-Council have followed in succession up to the present time.

This was a complete departure from the policy of the governments which had been in power since 1898. It came about as a result of the depression. Times were bad, not only in Northern Ontario, but all over the world. By 1932, world trade had shrunk to 32% of what it had been in the lush 1920s. All countries, without exception, were affected and were looking for trade. A few shiploads of pulpwood brought into the east American coastal ports and to Lake Erie paper mill centres from Europe were all that was needed to frighten the Department of Crown Lands into taking some drastic action, particularly when pressing demands were being made on the government by Lakehead timbermen to amend the law in order to permit shipments of pulpwood to American mills from Crown lands. Once you break a precedent, according to tradition it becomes difficult to restore things as they were. The lifting of the embargo at that time has been the source of great losses to Northern and Northwestern Ontario.

C.W. Cox, M.P.P. for Port Arthur, who had headed the delegation and claimed the credit for having this Order-in-Council passed, had this to say on the floor of the Legislature a year later.

> I wish to remind you of the conditions which existed in the pulpwood industry last spring. There was a great demand from the United States for exportable wood, which was being mostly supplied from Europe. Our woodsmen were idle and our tremendous forest resources were inactive, owing to government regulations which prohibited the export of wood to supply the demand. You will remember that I advocated a change in policy and that representations were made to the government which resulted in an Order-in-Council removing the export restrictions and making

available sufficient wood to meet the demand. This action had the
effect of marketing about 60,000 cords of wood, which employed
about 3,000 men for the summer months, to the enhancement of
the provincial treasury and the benefit of the industry. The Prime
Minister and the cabinet are to be congratulated upon the quick-
ness with which they grasped the situation and acted, and upon
the beneficial results of their action.

The sluice gates had been opened. One thing can be affirmed: C.W.
Cox must have had the support of all the timber contractors of the district when this special Order-in-Council was enacted.

Paradoxically, this Order-in-Council came into effect just when economic conditions were showing signs of improvement. The recovery,
however, was to be slow. Charles McKay, L.L.D., in the introduction
to his book *Extraordinary Popular Delusions*, stated, "Men, it has been
well said, think in herds; it will be seen that they go mad in herds, while
they only recover their senses slowly, and one by one." It took another
Great War to bring prosperity back to the North American continent.

Exports of pulpwood to the U.S. helped alleviate immediate economic woes. But the long-term goal remained to build mills in Ontario.
Here the Liberals adopted a new and radical approach. There can be no
doubt that the new Hepburn government in 1934 had inherited from the
previous Conservative administration a system heavily laden with political patronage and political consideration of every sort. Practically
all the fine timber stands of Northwestern Ontario had been conceded to
political friends and timber speculators. Though their intentions may
have been prompted by a desire for mills, few were built. Too many
sought only to make quick and spectacular profits from their timber
concessions. The disease was serious, and a method of surgery had to
be found to alleviate the pains of the patient. Whatever faults Mitch
Hepburn may have had, he did not lack courage. Nervous in disposition, he was quick to decide what surgical instruments were needed and
what anaesthetic to use. In 1936, his government passed the Forest
Resources Regulation Act, which empowered his Minister of Lands and
Forests to re-allocate timber lands to whomever he thought likely to
build either pulp or paper mills. The most unfortunate part of the bill
was that it entrusted the minster of Lands and Forests with absolute
authority.

This Act was the very essence of dictatorship and while the methods used for re-allocation of the timber wealth of the province might

have been desirable to the administration under such authority, they were bound to have serious consequences. The man entrusted with these new powers, the new Minister of Lands and Forests, was the member for Kenora, Peter Heenan.

Heenan's only experience in timber operations was as a locomotive engineer on the C.P.R. running out of Kenora to Ignace, and then to Winnipeg. He had seen plenty of trees along the roadbed and had hauled carloads of pulpwood to either Kenora or Dryden, but had never worked in the bush. He was, however, a real politician. As a Labour representative in a Farmer-Labour coalition in 1920, he had become heavily involved in a huge and controversial pulpwood concession. This concession of 3,000 square miles, granted by the Drury administration to the Backus Brooks Company, aroused considerable criticism among newspapers, especially as Backus had already obtained 1,800 square miles of timber land in 1912. It was felt that the government had given away an empire of forest lands, and the name of Peter Heenan became closely associated with the grant. At least he was enough of a politician to claim all the credit for it. He gave an interview to the Lakehead newspapers, setting forth the advantages the town of Kenora and the district would derive from the construction of a large paper mill. When the news reporters asked him "How did you do it?," he replied "Oh, that is my secret." E.W. Backus had been powerful for some years at the foot of the throne in Toronto. He never spared expenses in obtaining the services of skilled lobbyists to look after the interests of his various organizations. Be that as it may, Peter Heenan was basking in the glory of this achievement. Unheard of before, except in and about Kenora and Ignace, he rose to the stature of a provincial figure, and from then on his rise was fairly rapid. He returned to his former love and ran for the federal house in the next general election of 1926 as a Liberal candidate. Despite having practically no organization to begin with, he easily carried the constituency and was made Minister of Labour in the King administration. When the Liberals went down to defeat in 1930, Heenan was re-elected to the House of Commons, but he resigned in 1934 to re-enter the provincial field and take over as Minister of Lands and Forests.

The year 1937 was a year long to be remembered in the annals of the Lands and Forests Department, by the people of Northwestern Ontario in particular. No fewer than eight contracts were negotiated by Heenan with various companies, each undertaking to erect a mill in return for timber concessions. These firms included the famous Lake Sulphite Company, the General Timber Company, Pulpwood Supply Company, Huron Forest Products, Soo Pulp Products, English River Pulp and Paper Company, Vermillion River Pulp Company, and the Western Pulp and Paper Company. Under the provisions of the Forest Resources Regulation Act, the minister was not obliged to call for tenders; all these contracts were signed privately.

The first contract, that involving the Lake Sulphite Company, had special significance to the merchants of Northwestern Ontario. Construction of a new plant in Red Rock began in early April 1937, and before the following spring, with the plant still incomplete, the company was in bankruptcy. This promotion had been headed by R.O. Sweezey, an investment banker from Montreal who was familiar with the forest reserves of the Canadian Lakehead. In 1917, he had looked over the Black Sturgeon and Pic River limits, and in 1936 was acting in an advisory capacity to the president of the Great Lakes Paper Company of Fort William. During that year, he spent a great deal of time in and about the Canadian Lakehead where he accumulated additional knowledge of general conditions in Northwestern Ontario. He then decided to organize a bleached sulphite mill in either in Northwestern Ontario or Northern Quebec. Since most of his financial friends were located in Montreal, he preferred to get concessions from the Quebec government and to build a mill there. Accordingly, he approached the Honourable Maurice Duplessis, then Premier of Quebec with a proposition to erect a mill if his government would grant some concession from Crown lands. When Duplessis refused to entertain any project until, as he had stated shortly before to another promoter, the pulp and paper industry was out of the woods, so to speak, Sweezey decided to see what could be done with the Ontario government. In a brief meeting with the Premier of Ontario, a deal was mapped out. A contract was then entered into between Sweezey, on behalf of his projected company, and Peter Heenan, the all-powerful Minister of Lands and Forests, by which Sweezey agreed to build a bleached sulphite mill at Red Rock. In return, he received one of the finest timber concessions ever given to a promoter.

The situation at the Lakehead was not good. The recovery from depression was slower than had been expected. There was little construction of new homes and public buildings in either Port Arthur or Fort William. Men were looking for work, and merchants for trade. It was felt at the time that the construction of this new mill in the district was a blessing, and until it went broke, the company employed a great many men, and bought considerable supplies and equipment. There was some real activity in and about Nipigon and Red Rock, but it proved to be only temporary and very costly to the region's merchants and manufacturers. This enterprise was unquestionably badly timed. When they began to construct their mill, bleached sulphite commanded a good price. Before the end of the year, the price had dropped 30%. The market was already over-saturated. The project was only partly financed. Had this company been well provided with capital, there is no doubt that they could have carried on, even if the operation had not been profitable for two or three years. In time, it would have proven to be a good investment and a great benefit to the people of the region. How the Department of Lands and Forests ever permitted a project of this kind to go ahead without assurance that they were capable of carrying it out, is still a mystery.

There was, of course, an election that year. R.O. Sweezey, a highly intelligent individual well versed in political craftsmanship, was aware of that. He had had a great deal of experience in dealing with politicians when he obtained his charter to develop power for the famous Beauharnois Corporations of which he was president. Speaking in Winnipeg in 1929, when this great project was being started, he had stated that he knew from the beginning that the development of power along the St. Lawrence (through the building of a huge canal, which could in time be used as a part of the deep waterway system), was not so much an engineering problem as a political one; he was satisfied that he had succeeded in mastering the greater of the two problems, and that his vast undertaking was on its way to a successful conclusion. A year and half later the deal was investigated by a parliamentary committee at Ottawa. It was proven that R.O. Sweezey had been most generous to friends both in government, and in opposition. When pressed by one of the committee members as to why he had handed out such generous political contributions, he replied that he was told that it was the proper thing to do. The Beauharnois project was one of great magnitude. Its engineering planning was faultless. The cost of its construction was less than had been estimated by the engineers of this project. As R.O. Sweezey had stated in Winnipeg, it was not an engineering, but a political problem, and it was this problem that eventually defeated him.

The Lake Sulphite project must have appeared insignificant to him in comparison with the Beauharnois power development. But Sweezey had three distinct problems to deal with at Red Rock: political, financial and economic. He was not responsible for the third except in bad timing, which showed lack of judgment, but he can be held accountable for the second one, since his undertaking was not sufficiently financed. As to the political problem, it can be assumed that he was fairly generous in handling it. At any rate, during the 1937 election, attractive billboards, appealing advertisement and radio addresses urging the people to return Hepburn were in evidence all over the province. The party was well financed for that election. But that was not all. At the height of the campaign, Peter Heenan addressed a banquet in Fort William. He had been referred to by the chairman, and in advertisements through the local press for two or three days previously as the greatest authority on our forest resources in the history of Northwestern Ontario. During his address he announced that no fewer than eight paper mills were going to be built in Northwestern Ontario involving an expenditure of approximately $44,000,000: the Lake Sulphite Plant at Nipigon, at an estimated cost of $9,000,000; new mills at Fort William and Kenora, to cost about $5,000,000 each; a new mill at Michipicoten, $5,000,000; a new mill at Blind River, $4,000,000; and new mills at Longlac and Marathon (Big Pic River), $2,000,000 each. Extensions planned or underway included, he said, Thunder Bay Paper mill at Port Arthur, estimated cost, $1,000,000; Great Lakes Paper at Fort William, about $2,000,000; Abitibi Mill at Fort William, $1,000,000; Fort Frances mill, $1,000,000; and other Abitibi mills, about $6,000,000.

He completely stole the show that night at the banquet. C.W. Cox, who was seeking re-election for the constituency of Port Arthur, had to play second fiddle. The people of Northwestern Ontario had never, in all their experience, been visited by such a Santa Claus. Heenan further claimed to have cured unemployment in his own constituency. For the next ten days, until the election, anybody who traveled along the Trans-Canada Highway between Thunder Bay and the Manitoba border, or over the Heenan Highway to Fort Frances, could see all the unemployed at work on the road, some of them with rakes in their hands and most of them with overcoats on. As a matter of fact, there were a number of fires along the road where some of these loyal supporters could keep warm. With such promises of gifts right and left and a sure cure for unemployment, Heenan carried Northwestern Ontario and Mitch Hepburn won a new mandate; he appeared to be safe for the next five years.

It was becoming evident, however, that the ill-feeling which had existed for some time between Heenan and fellow Liberal C.W. Cox, member for the Port Arthur constituency, was developing into a feud. Not many people on the outside knew how serious it was until one evening when Peter Heenan addressed a meeting at Schreiber in support of Cox. It so happened that the Conservative candidate for the Port Arthur riding, George Wardrope, whom Heenan knew intimately, was in that town. Peter, who was fond of good tobacco, good food and good liquor, may have imbibed too much of the latter for he became loquacious, and said to Wardrope, as he was getting on board a train: "Say, George, if you don't lick that black S.O.B I'll come back after the election and give you the worst licking you ever had." Cox at that time was an attractive looking man of dark complexion and jet black hair.

Of the eight agreements already referred to between the Crown and the promoters, all but three defaulted. The assets of the Lake Sulphite Company were purchased by the Brompton Pulp and Paper Company and eventually taken over by the St. Lawrence Corporation Ltd., who completed the plant, installed necessary machinery and equipment to manufacture kraft paper and newsprint, and extended the townsite. The other two organizations, who have more than met their obligations to the Crown, were the Marathon Paper Mills of Canada Ltd. and Kimberly-Clark Pulp & Paper Company Ltd. Let us turn first to the Marathon company.

It has already been recorded that the Alexander-Stewart Lumber Company of Rothschild, Wisconsin, had purchased a quarter interest held by Herman Finger in the Pigeon River Lumber Company in Port Arthur in the autumn of 1904. This share had been divided between Alexander Stewart and Walter Alexander. It was a few years afterwards that they, in partnership with C.C. Yawkey and D.C. Everest, built a paper mill at Rothschild. This was the origin of the Marathon Company. They had been interested in logging and sawmill operations on the North Shore almost continuously from 1904 until the old Pigeon River Lumber Company abandoned its operations in the early 1920s. Ben Alexander, a son of Walter Alexander, became a partner of E.E. Johnson when the latter organized the Pigeon Timber Company at about this time. The officials of the Marathon Company were therefore quite familiar with the timber resources of the North Shore.

Accordingly, in 1935 they suggested to Al Johnson (now Colonel Johnson), who was associated then with his brother E.E. Johnson, that he make a study of the north shore of Lake Superior and find out if there were any sites available on which a pulp mill and a townsite could be constructed. In his report, Al Johnson suggested their operations be located in the vicinity of the Big Pic River. In 1936 Peter Heenan, Minister of Lands and Forests, gave permission to the Marathon Company to cut pulpwood for export on the Black River limit. Accordingly, Marathon formed the General Timber Company Ltd. for this purpose, with Al Johnson as general manager. In 1937, however, the government, through Heenan, re-allocated the Black River limit to the Ontario Paper Company which began operating a rossing mill near that point. The General Timber Company was given cutting rights on the lower Big Pic River, but no concession was yet allocated to them. During the same year, the Marathon Company made an agreement with the Crown, through its contractor, General Timber, which provided for construction of a pulp mill in the District of Thunder Bay. They accordingly were awarded the concession they had wanted from the start, the area known as the Big Pic River watershed. Operations started immediately with the wood being driven down the Pic River and towed in boom across Lake Superior to Ashland, Wisconsin. It is worth remembering that the manufacturing clause of the Pulpwood Act had been suspended making pulpwood from Crown lands exportable.

In 1938 the Marathon Company bought out the interests of the stockholders of the General Timber Company which, therefore, became a wholly-owned subsidiary. In 1943, in fulfillment of its agreement to build a mill, the Marathon Company was awarded an additional timber limit known as the Nagagami Concession, which would provide additional pulpwood for the domestic mill they intended to build at Marathon (then Peninsula) and also sufficient pulpwood for both the parent company in Wisconsin and the Rhinelander Paper Company which had helped finance the Nagagami Concession. The construction of the mill at Marathon began on April 1, 1944 and it was in operation two years later. The pulpwood taken out from the Nagagami limits — a total of 100,000 cords — was shipped by rail on the Algoma Central to Sault Ste. Marie and then to destination by American railways. All the wood cut from the Big Pic River basin was to be utilized locally at the Marathon mill.

The mill at Marathon grew to be one of the most modern mills in Ontario; the townsite is particularly attractive. It was later controlled by the American Can Corporation. Although handicapped by a shortage of labour and difficulties in obtaining building material, machinery and other equipment, the company persisted and were in operation by October, 1946. By 1960, the mill and townsite employed 700 men with an annual payroll of $1,250,000. In their woods operations they employed 1,500 men with an annual estimated payroll of $3,500,000. John Stevens Jr. was president of the Canadian company at that time and Paul Klinestiver was vice-president in charge of woods operations. In addition to their own office staff at Marathon, they had a large number of office employees and woods executives in their Port Arthur office.

The Toronto-based Kimberly-Clark Company originated in the Pulpwood Supply Company, an exporting firm which enjoyed an unusual franchise of pulpwood limits and waterways on which the government had spent a considerable amount of money (by diverting the water from Long Lac to Lake Superior). This firm was a subsidiary of the Kimberly-Clark Corporation of Neenah, Wisconsin. The president was Col. G. Parker with Charles H. Sage vice-president, and H.S. Craig as general manager. They constructed a large mill and townsite at Terrace Bay, some nine miles east of Schreiber. It was one of the most modern plants on the continent, as well as one of the most beautiful town sites to be seen anywhere, both of which represents a capital far in excess of what had been undertaken at the time their agreement was signed in 1937. Their activities in this district were a contributing factor to the welfare of this section of Northwestern Ontario.

The history of the Kimberly-Clark enterprises in Canada is fascinating. They, in conjunction with the *New York Times*, built a small mill at Kapuskasing in 1920. Six years later they began the construction of a modern paper mill and a townsite. This project had originally been planned prior to the 1920s by S.A. Mundy and Associates of Bradford, Pennsylvania. Elihu Stewart of Ottawa, who had been associated as chief forester for the Dominion government, was in charge of forestry planning. Mundy and Associates had obtained 1,740 square miles of timber lands including the power rights on Sturgeon Falls and Beaver Falls on the Kapuskasing River. Conditions were such, however, that the Bradford syndicate was unable to carry their project to completion. Not long after, George F. Hardy, a leading consultant engineer of New York, intimated to the officials of Kimberly-Clark that the New York Times Publishing Company might be interested in pursuing the project. An agreement was soon reached and a new company, the Spruce Falls Power & Paper Company of Ontario was organized to carry out the scheme. Having acquired all the timber and water rights from the Mundy interests, Spruce Falls began construction of a small mill in 1920. Some few years later, J.H. Black was appointed general manager, with Egerton S. Noble and George Barber, of North Bay and Toronto respectively, as executive assistants; Noble later took over as general manager. In 1960, the mill was producing 650 tons of newsprint and 250 tons of sulphite pulp per day. In addition to this, they operated a railway 75 miles distant to Smoky Falls, on which a large power project has been developed. This vast enterprise of Kimberly-Clark at Kapuskasing, developed in conjunction with the *New York Times*, can be reckoned an outstanding achievement.

Kimberly-Clark, like many firms of the last century, had a humble origin. In 1848 the state of Wisconsin was formed. Lands were opened to settlement and large-scale immigration followed. A great number of settlers and artisans came from Holland and Germany (both of which countries were undergoing a period of revolution); others came from the Scandinavian countries and a smaller number from Eastern Canada. It was about this time that John Arpin, father of D.J. Arpin, left Quebec with his brother to take up timber land near what is now Wisconsin Rapids, and in time established the Arpin Lumber Company. The eldest son, as we have seen, founded the Pigeon River Lumber Company which played such an important part in the district for the first two decades of the twentieth century. A considerable number of enterprising young men came from the Eastern States, some of them traders and others experienced lumber operators. Among these were J.A. Kimberly, C.B. Clark, F.C. Shattuck and Havilah Babcock; all in their early years when they went into partnership. Kimberly was 34, Babock 35, Clark 28 and Shattuck 34. Two of them went into partnership in a general store. C.B. Clark became a partner in a retail hardware business and F.C. Shattuck developed a wholesale business. In a matter of a few years the four between them had saved $30,000. It was with this capital that they started manufacturing paper.

The firm was called Kimberly-Clark and they began operating their mill in 1872. From this small capital, their business expanded rapidly. It developed from the small plant at Neenah through progressive, if cautious management, into a leader among all the mills of the Fox River Valley. Finally they began to expand outside the state of Wisconsin, having vast interests in Kapuskasing and at Terrace Bay, Ontario. The Kimberly-Clark Corporation of Neenah employed 6,000 people in 1960, with an annual payroll of $16,396,685. In 1945, there were 4,000 stock-holders, many of whom were employees of the firm. The company paid dividends that year amounting to $1,660,429. J.A. Kimberly, last of the original four partners, died in 1928 at 90 years of age. He was succeeded as president by F.J. Sensenbrenner, who retired in 1942.

It may be well now to return to the late 1930s and visualize what was taking place from the return of the Hepburn government to power until the early 1940s, when the feud between C.W. Cox and Peter Heenan, Minister of Lands and Forests, was brought out on the floor of the legislature. As has been stated, none of the mills promised the people of Northwestern Ontario in election speeches were being built, except the Lake Sulphite, which had gone bankrupt in 1938. Some of the grantees of these limits were, at the same time, taking every advantage of their contract to export pulpwood to their mills in Wisconsin and Pennsylvania, without making any effort to live up to the conditions attached to their agreement with the Crown through the Minister of Lands and Forests. It took a change of government to bring about a solution. The grantees had either to build mills or forfeit their concessions.

The Premier of Ontario, Mitchell F. Hepburn, who had been attacking Conservatives when he assumed control of Queen's Park in 1934, was now shooting at federal Liberals and, in particular, at the Prime Minister of Canada. As a matter of fact, Hepburn had already begun to criticize the federal party before the 1937 election, but only in a perfunctory manner. After that date, however, his campaign against McKenzie King became an obsession, an intrusion into national affairs that led him to neglect the leadership of his party in Ontario and the responsibility of governing the province through legislative means. The next few years, therefore, were to be a record of aimless administration. If we look back over the history of Ontario, it is discovered that the premiers who rendered the greatest service were those who devoted their entire time to the administration of the province. It is a full-time job, requiring skill, vision and character. Mitchell Hepburn had an opportunity rarely given a man and might have gone down in history as one of Ontario's greatest leaders had he stuck to administering the affairs of the province. Instead he lost the confidence of his own cabinet; eventually, the people themselves settled the matter when they elected George A. Drew with a mandate to carry out aggressive government policies.

After 1940, the elements of discord within the Liberal cabinet were not long in wrecking the administration. Outstanding among these signs

was a speech made by Charles W. Cox in the legislative assembly on
March 18, 1941 lambasting the Department of Lands and Forests and
Peter Heenan; a sad exhibition of disunity within the Liberal party.. It is
well to remember that it was Cox who had so strongly advocated the
Order-in-Council permitting the exportation of pulpwood from Crown
lands. The Port Arthur member had by this time completely reversed
his former stand. Whether this change of heart and attitude towards
government policy was brought about by his personal feud with Heenan,
or from the actual facts evident in the case, only the former member for
Port Arthur would know. This is what C.W. Cox said seven years after
he had complimented the government on having lifted the embargo on
the exportation of pulpwood to American mills:

> My personal opinion is that the present policy is not only ruinous
> to the district which I have the honour of representing, but will
> have a far reaching and detrimental effect on the province and the
> dominion. The district I represent occupies an important position
> in our lumbering industry. Its geographical location affords low
> transportation cost by water to some of this continent's largest
> markets. Its timber resources are extensive, and, for the most
> part, they are tributary to the north shore of Lake Superior. The
> district's importance is further attested in the fact that approxi-
> mately one-third of the total revenue obtained by the Department
> of Lands and Forests of the province of Ontario is derived from
> the Port Arthur office. I think I am safe in saying that about 90%
> of the pulpwood exported from the Crown lands in Ontario is
> likewise from the same territory.
>
> You may have heard of the re-allocation of timber areas. This has
> been done in a manner which gives the exporting mills, and those
> supplying the export market, a very distinct advantage over
> Canadian mills. As an example: our local Abitibi Company is
> forced to go 200 miles from their mill in order to secure their
> supply of wood, necessitating driving and towing operations
> which require large expenditures of money, while foreign mills
> and those operating for them, are given limits on the shore, and,
> in some cases, locations where improvements have been made at
> public expense.
>
> We have at the Lakehead a number of men representing our pulp
> and paper industry, an industry having an investment of approxi-
> mately $30,000,000, constituting the principal business of our
> district. These men realize that the present policy is ruinous, but

did not dare appear at the timber probe for fear of reprisals. As a matter of fact, one man who represents a very large concern expressed a desire to give evidence and was told he had better 'stay away'. I am to some extent participating in the very policy I condemn, having been associated with the lumbering business for the past 25 years. Under the present policy, any operator is compelled to be a party to it, if he would stay in business.

It was a serious charge against the administration, but one that had no telling effects; Cox, as he himself admitted, had no credibility on this issue. It was well known in his own constituency of Port Arthur that he had participated in the period of piracy or privateering. He had exported pulpwood from his patented timber lots which had been staked as mining locations. Considerable piling timbers had been cut off these properties to be used in elevator construction. Nor had he hesitated to trespass across lines. He had simply followed the pattern of the time. He had been no worse, but no better, than the timber buccaneers of that period. Finally, while he was railing against the government policy which permitted the exportation of pulpwood from Crown lands to American mills, he was also participating in that very policy! In and out of the legislature, and in his own constituency, Cox's verbal attacks were taken no more seriously than those of an advocate of prohibition who held, or was suspected of holding, shares in a brewing company.

Cox had expressed the hope in his address to the assembly that something constructive would develop from the timber enquiry then underway, but he was becoming extremely doubtful. In 1941, a select committee was appointed to investigate, inquire and report upon all matters pertaining to administration, licensing, sale, supervision and conservation of natural resources by the Department of Lands and Forests. The committee consisted of J.M. Cooper, chairman, H.C. Nixon, A.L. Elliott, F.R. Oliver, W.G. Nixon and Peter Heenan, representing the government; George A. Drew, Frank Spence and Harold Welsh, represented the Conservative opposition. Two reports were submitted — a majority report by the members of the administration, and a minority report by the opposition. The first had one purpose in view: the whitewashing of the administration. The second was a denunciation of the abuses that had crept in under the administration of Peter Heenan.

The minority report was well thought out, clearly elucidated and its recommendations suitable to the cure of the disease. It quoted from two paragraphs of the Riddell-Latchford Royal Commission submitted to the House in 1922, as follows:

> We are of opinion that no office, minister or otherwise, should
> have the power to grant rights over large areas of the public
> domain at will, without regard to regulation; that power was never
> contemplated by the statute, it does not at present exist, and
> should not be given to any individual. Such an arbitrary power,
> subject to no control, is obviously open to abuse.

We found that in many cases there is no departmental record of
applications that were afterwards granted; in some cases, there

---

Key to Map 5 Selected Timber Limits, Mining Locations

1. **John Nesbitt & Co.** (Sarnia), patented properties, mouth of Pic River
2. **Hazelwood & Whalen**, patented properties, Pic River to Black Bay, 1898
3. **J.T Emerson, Joe Brimstone, James Meek, Robert E. Whiteside**, along Willow River, 14 miles east of Pic River
4. **J. St. Jacques and C.E. Smith**, old mining lots, Prince Bay, Jarvis Bay, and Cloud Bay, 1907-12
5. **E.A. Carpenter**, mining properties, Cloud Bay and Prince Bay, 1890s-1900
6. **C.E. "Haywire" Smith**, mining locations, Prince and Cloud Bays. Founded Central Contracting Co. with operations on patented mining lots mostly along Black Bay penninusla
7. **Port Arthur Pulp and Paper Co.**, Sibley penninula
8. **Great Lakes Paper Co.**, Pic River limits, 1922
9. **Marathon Pulp and Paper Co.** mill
10. **Kimberly-Clark**, Terrace Bay mill
11. **James Hourigan Co.** limits, Black Bay penninsula
12. **Thomas Marks**, 1895-96
13. **Hazelwood and Whalen** patented properties, 1898
14. **North Shore Timber Co. of Port Arthur**, Charles McCarthy, patented properties at mouth of Black Sturgeon River
15. **Lake Superior Timber Co.**, J.T. Miller (Detroit), cutting rights from mining locations of Montreal Mining Co. and Ontario Land and Mining Co., St. Ignace Island, Simpson Island and along Nipigon straits, 1902
16. **Northern Island Pulpwood Co.**, Walter H. Russell, acquired properties of Lake Superior Timber Co., veteran's land grants, along Trout Creek near Nipigon and on the banks of Black Stugeon, 1907. Rights sold to Newago Timber, 1914
17. **Russell Timber Co.**, Black Sturgeon River, 1911
18. **C.W. Cox**, patented timber lots at Shesheeb Bay, Black Bay penninsula, 1912-13
19. **James Hourigan Ltd.** timber limits along shore of Shesheeb Bay, assumed by C.W. Cox in 1920.
20. **Edward E. "Eddie" Johnson** timber limits on inside of Black Bay penninsula, 1920s
21. **Nipigon Fibre and Paper Mill Ltd.**
22. **Lake Sulphite Co.**, Red Rock
23. **Great Lakes Paper Co.** timber limits, Black Sturgeon, 1922

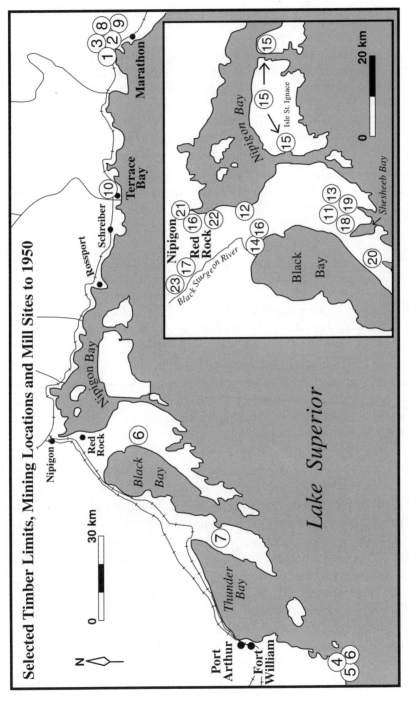

Map 5 Selected Timber Limits, Mining Locations and Mill Sites to 1950

was conflicting evidence as to these. We think this an improper practice. All applications to the department or any officer thereof, for a concession of or right or privilege in any part of the property of the province, should be in writing, should be entered as received and should be placed on the public files. If at any time an oral application be made, a full memorandum of the application should be prepared forthwith and duly filed. It is unwise to leave matters of importance to fallible recollection, particularly matters in respect of which there is a trust for the people.

The authors of the minority report further stated that they were in entire accordance with the Riddell-Latchford recommendation and believed that it should have been strictly observed. "It was followed by the present minister in scarcely any single instance," they wrote,

> although properties worth scores of millions of dollars were being dealt with. The evidence is clear that matters of the utmost importance were left to extremely fallible recollection, and in most cases, there were no memoranda, no surveys upon which to base values, and it was utterly impossible to ascertain from the minister upon what grounds he had reached his decision to grant very large and extremely valuable areas to companies which had no apparent assets except the property thus acquired.

It has been stated that party labels are only skin deep. It may be added that good intentions are very shallow. It is a sad commentary on human character that men of probity and good living habits who would not think for a moment of taking advantage of their fellow men in any deals, but when confronted with the temptation of acquiring a timber holding from the government, drop all sense of proportion and even go absolutely contrary to some of the policies they have advocated or the reports they have supported. It was discovered some time after the minority report that Frank Spence, one of the signatories, was the beneficiary of a timber limit given to him over the head of the deputy minister, through the influence of Mitch Hepburn, the Liberal leader. Hepburn was dispensing a favour to a Conservative stalwart from one of the Lakehead constituencies.

Unfortunately, to a large extent this work has been a recital of lobbying, political contribution, high-grading, infractions, buccaneering and privateering, which meant defrauding or depriving the Crown of just revenues. This could not well be otherwise, since, for the preceding thirty five years, the Department of Lands and Forests had been made a political football. During all that time, for an operator, speculator or

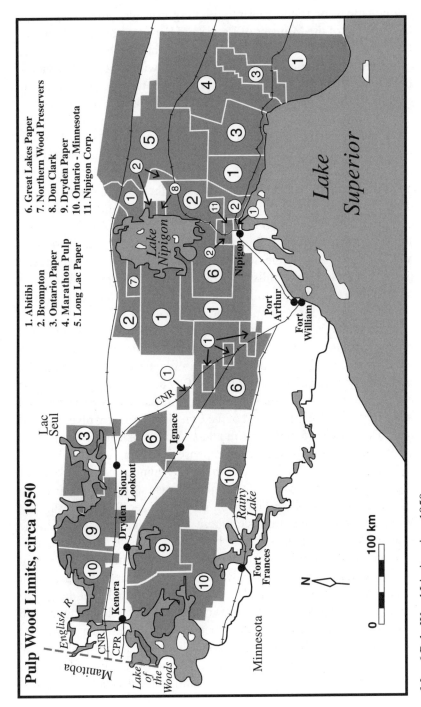

Map 6 Pulp Wood Limits, circa 1950

promoter to be on friendly terms with the government of the day and to have the ear of the Minister of Lands and Forests, was equivalent to a franchise. In and about the Lakehead district a timber operator was estimated in proportion to the pull that he had with either the Premier or the Minister of Lands and Forests; political patronage was always an operator's first consideration. Nowhere else in Ontario did timber barons take such advantage of Crown lands.

It may be well to remember what took place in Michigan, Wisconsin and Minnesota which at one time possessed possibly the finest stands of white and norway pine on the face of the earth, in addition to vast areas of spruce and other conifers. Most of these vast areas have been denuded of their timber and their forest resources dissipated. What the exploiters had failed to do in denuding the forests of the three states, forest fires accomplished. The people of two of these states, Wisconsin and Michigan, have witnessed the irrational spectacle of large investments in pulp and paper mills when there were insufficient timber stands of their own to keep their mills in operation and they had to depend on Canadian forests for their pulpwood supply.

The methods which some of the U.S. robber barons used to get timber lands in all three states in the last half of the nineteenth century were not vastly different from those later used in Northwestern Ontario, particularly in and around the Thunder Bay district. If there was some difference, it was in the amount, and in the areas. The Americans did it in a wholesale way, while most of the Canadian Timber Wolves seem to have been satisfied to remain as retailers. State legislatures were packed by representatives whose election expenses had been furnished by lumber barons. Some of the magnates in the U.S. lumber industry even rose high in government, where they could exercise still greater control. In the early 1880s, the elder Frederick Weyerhaeuser, founder of the vast Weyerhaeuser Corporation, said there was enough pine timber in the state of Wisconsin to last at least 100 years. His company had extensive limits in the northern part of the states at that time. By the end of the century, all the pine stands were destroyed. The last of the Great Lakes states to have any stands of virgin pine forests left at the beginning of the century was Minnesota whose northern timber lands had been saved to some extent by their inaccessibility. But when the forests of Wisconsin and Michigan gave out, there was a mad rush to gain control of the white and norway pines of Minnesota. The grabbers had even improved their methods and the land office was helpless against their raids. Cheques were unearthed showing payments up to $6,000 for "greasing," the term generally applied to bribery. Timber lands had been obtained during that boom period in Northern Minnesota by wilful misrepresentation, and by every device available to land speculators to defraud the state of revenue from stumpage dues. In ten years, no fewer than 34,000 investigations into fraudulent entries had been ordered, but, unfortunately, public opinion supported the practice.

The same sort of corruption was evident in Ontario. An entry entitled "grease," representing an amount of $40,000.00, was produced as an exhibit at the timber investigation held by a committee of the legislature of Ontario. When requested by the Honorable George A. Drew, then leader of the opposition, to explain the significance of that entry, the president of the Company sharply retorted "it means exactly what it states." E.E. Johnson, who did not part with his money easily, was known to be generous at election times to candidates of his choice. This must be accepted as a reason why all the Royal investigations held in Ontario, whose reports have been submitted and recommendations made, are just so many records in the files of the Department of Lands and Forests. The department was so "hogtied" by political patronage and political consideration that reform was well nigh impossible.

All across the north, the forests are in trouble. Minnesota's experience should provide us with a clear lesson. In 1899, the saw mills in that state actually manufactured 2,342,000,000 feet of lumber and, by the beginning of the present century, the Duluth and Superior harbours alone were shipping out 1,000,000,000 feet a year. In 1944, however, when lumber was so badly needed to carry out the war effort, and the American government was using all its resources to stimulate enterprising small mill owners, Minnesota produced only 250,000,000 feet of lumber. The sawmill industry, which had been a prime industry of that state in the years before 1920, completely disappeared. We in Northwestern Ontario have something to ponder. Our climatic conditions and geological formations are similar to those of Northern Minnesota, as they are in the entire territory bordering on Lake Superior. Forest lands do not lend themselves to rapid reproduction, particularly when they have been completely denuded.

In the early part of the twentieth century, Rainy River, Kenora and Keewatin were all prosperous sawmill towns. After 1910, however, the timber line receded dramatically in and about these towns. The Shevlin-Clark Company of Minneapolis which held vast pine limits on the Ontario side of Rainy Lake and the adjacent watershed, erected a large mill at Fort Frances, which was kept in operation until 1943 when their pine limits had been cut out. J.A. Mathieu, formerly general manager of the Shevlin-Clark mill, had acquired some timber limits of his own, and organized the J.A. Mathieu Company in 1922. He also acquired a sawmill two miles east of Fort Frances which was in continuous operation until 1954. This firm and the Patricia Lumber Company of Sioux Lookout were the only two substantial sawmill operations left in the area west of the Canadian Lakehead cities to the Manitoba boundary.

It is the same story around Georgian Bay and along the whole north shore of Lake Huron to Sault Ste. Marie. In 1904, mills operating in that region were cutting 600,000,000 feet of lumber per season; all but one completely disappeared and what were once prosperous communities became deserted ghost villages. The same thing has occurred in the Thunder Bay district.

Northern and Northwestern Ontario still have large stands of timber, but they are among the few remaining large forest areas on the American continent. These forests are the property of all the people of Ontario and they must he utilized to the best advantage and for the greater benefit of the people as a whole. The record of destruction of vast forest wealth in the northern U.S. and in Ontario makes a dismal story.

Are the Timber Wolves of tomorrow going to he less greedy and voracious than the timber operators have been in the past? Human nature has not changed much in the past 4,000 years. Let us hope that the officials of the Department of Lands and Forests will exercise more vigilance than in the past. There have been some positive signs. The *Globe and Mail* of June 24, 1947, featured an article on its front page regarding trespassing by timber operators in the province. In it they quoted Harold Scott, Minister of Lands and Forests who stated that "timber operators had been cutting from Crown lands without proper authority." The minister warned the "Timber Wolves," or trespassers, that if this illegal cutting does not stop, "action will be taken under the criminal code." "The Department," the minister said, "takes a serious view of the situation and cases reported will be made public, and the guilty parties prosecuted." He explained that the two broad types of illegal cutting reported were: cutting without authority of any kind (which actually is stealing Crown timber and is liable to a charge of theft); and the ruse of cutting over the line, or trespassing, an old method of acquiring timber without paying dues. The minister further warned that as soon as illegal cutting was discovered the district foresters would he instructed to place a seizure notice on the cut.

It was maintained at the beginning of this chapter that pulp and paper companies no longer just manufactured paper, but became financial conglomerates heavily influenced, if not controlled, by investment bankers. This has had an unfortunate impact on the forest resources of North America. Are the hard-boiled financial men and executives of these corporations going to be less greedy tomorrow than they were yesterday? The signs are not good. A recent investigation in New York was conducted by the Department of Justice of the United States into the newsprint industry. According to an article dated from Montreal and appearing in newspapers all over the United States and Canada at that time, it was alleged that the International Paper Company, one of the large paper firms operating in Canada, independent of its parent organization in the United States, was clearly a case of a cartel dodging its responsibilities through a series of incorporations in Canada. According to the same article, U.S. Congressman Cecil King of California told the House of Representatives that there was no greater proof of shady shenanigans within the Canadian newsprint trust than the fact that these companies were fighting to avoid producing their records. If they have nothing to hide, nothing to fear, then why do they go to great lengths in refusing information to the Grand Jury?

Combines, trusts and cartels, seek by whatever means possible, whether it be by price fixing or a division of territories, to restrict trade. They are the very antithesis of free enterprise; if their actions are not checked, another slump will inevitably follow. In the 1930s, Franklin Delano Roosevelt stated that the money changers have now fled from their high seats in the temple of our civilization. Were he alive two and a half decades later, he would be distressed to hear from Rabbi Israel M. Goldman of Providence, R.I., president of the Rabbinical Assembly of America, that the "economic royalists were again setting up their thrones in the halls of American Democracy." We have, therefore, progressed very little from where we started in this study. Timber Wolves still lurk in the forests of Northwestern Ontario and greedy executives of large newsprint corporations still threaten to undermine the very system that has given the North American continent its position in the world. Between 1925 and 1930, for example, the capacity of Canadian newsprint mills doubled, but the consumption of newsprint in the United

States increased only 20%. The International Paper Company generally took the initiative in establishing prices for newsprint in the prosperous 1920s through its associated subsidiaries here in Canada, and its leadership was usually accepted by other newsprint manufacturers. In fact, three corporations in Canada at that time produced practically half of the Canadian newsprint: the Consolidated Paper Corporation, the Abitibi Power and Paper Company and the Canadian-International Paper Company. What was the result of this intensified production? Namely the further depletion of Canadian forest resources and finally a terrible slump in newsprint prices which had reached a bottom by 1933 when newsprint was selling at $41 per ton.

The regulation of trade rests with the federal government. It is their duty to see that no combine, trust or cartel takes root. In the United States there is the Antitrust Law, called the Sherman Act; in Canada we have the Anti-Combine Act. The administration of this act in Canada rests with the Minister of Justice; likewise the Department of Justice in the United States has a department devoted to the safeguarding of the interests of the people. We must ever be on the alert. Another collapse of the newsprint industry would prove most disastrous, particularly in Northwestern Ontario.

The administration of our vast forest resources is primarily a provincial responsibility. The pulp and paper industry, however, is of such economic importance to Canada as a whole, that our federal government created a Department of Forestry to co-operate with the various provinces in accelerating means of forest protection, and sustain forest yield; in short, forest in perpetuity. To attain that objective will require team work between all concerned: government, industry and labour.

Epilogue

J.P. Bertrand's call for more progressive forest management and less political corruption has not fallen on deaf ears. This book, as an unpublished manuscript, has circulated within the scholarly community (in the form of unauthorized photocopies) for a number of years since his death in 1964 and has undoubtedly influenced many. If he were still alive, he would no doubt be heartened that his plea for sustainable yields from the forests of Northwestern Ontario is now the stated aim of all the major forest industries in operation and that considerably less is heard today about industry and government corruption.

Bertrand's main concerns, however, have not vanished. The issues of clear cutting and forest sustainability are still unresolved and new problems such as pollution and bio-diversity have been added. Governments, as well, continue to play politics with the province's forest resources. And are we really sure that the Timber Wolves are gone?

It should be remembered that J.P. Bertrand was not a tree-hugger whose opinion can be easily dismissed. He was a man with deep roots in the forest industry and a true appreciation for the lumber worker and the life he led. One even senses a grudging admiration for the timber pirates themselves, for their audacity and boldness. But, most of all, J.P. Bertrand loved and respected the forest and the region he called home. He recognized, as few others have, that the future of Northwestern Ontario, the land he cherished, depends on the vitality of its forests.

Thorold J. Tronrud
August, 1997

Appendix 1

Biographical Index

Aldrich, Sherwood (1868-1927) was born in Riverhead, Long Island, New York and received a law degree from New York University in 1889. His business was resource development, primarily in mining. In 1890 he was involved in a Colorado mine. He lived in Great Neck, New York.

Alger, General Russell Alexander (1836-1907) was born in Lafayette township, Ohio. He worked as a teacher before being called to the bar in 1858. He practiced law in Cleveland. In 1860 Alger moved to Michigan and entered the lumber business. His business activities were interrupted by the American Civil War where he served as an officer. After the war he returned to the lumber business and headed the Alger Smith Company and the Manistique Lumber Company. He was engaged in logging operations in Michigan, Wisconsin and Minnesota and extended operations over the border across the Pigeon River into Ontario. In 1902 Alger's timber and logging equipment was taken over by the Pigeon River Lumber Company.

Angus, Richard Bladworth (1831-1922) was born in Bathgate, Scotland. After a successful career with an English bank he joined the Bank of Montreal in 1857 and later became an agent for a Chicago bank and a New York financial agency. In 1879 he resigned to become a director to the Bank of Montreal and eventually became it's president in 1910. With Lord Strathcona and Lord Mountstephen, Angus and others managed the St. Paul, Minnesota and Manitoulin Railway and formed the syndicate that constructed the Canadian Pacific Railway.

Anson, Francis H. was born in Niles, Michigan, but began his business career in Minneapolis where he was the manager of the foreign department of Ogilvie Flour Mills Company. In 1904 he was appointed general superintendent of the company that had its head office in Montreal. Anson was also director of the Dominion Box and Package Company and the Abitibi Pulp and Paper Company.

Arpin, Daniel J. and William Scott interested F.P. Hixon and Herman Finger in organizing the Pigeon River Lumber Company. Arpin was the president of the Wisconsin lumber firm and purchased the Graham and Horne sawmill in 1901. Herman Finger was his general manger. The company held vast timber limits along the Arrow and Pigeon River wartersheds. The following year they opened a modern sawmill at Port Arthur, the region's first "giant sawmill." The most modern construction methods were used and state of the art operating equipment was installed. The Pigeon River Lumber Company took timber from both sides of the Pigeon River and drove the logs to the mouth of the Pigeon River and towed them to the mill. The lumber output including lath and shingles was between 20 and 30 million board feet a year and was 35 million board feet for several years.

Backus, Edward Wellington (1860-1934), although born in Jamestown, New York, grew up in Red Wing, Minnesota. Backus acquired some training at the University of Minneapolis, but did not complete his courses there. In 1882 he joined a Minneapolis lumber company as a bookkeeper. In 1885 Backus bought the firm and established E.W. Backus Company. The E.W. Backus Lumber Company was formed in 1894 and, in 1899, William Brooks became his partner and the Backus-Brooks Company began. Backus was a businessman of great imagination and some daring and was the developer of Rainy River country. His businesses included the Koochiching Company, the International Lumber Company which did the logging and railroading for his Ontario and Minnesota Paper Company, and a large lumber mill at International Falls (1909-1932). Backus organized and was president of Kenora Paper Mills, Kenora Development Company, International Telephone, and the Minnesota, Dakota and Western Railway, and he played a role in the early development of the Great Lakes Paper Company. Backus was known as a man of great energy and appetite and he consumed a prodigious quantity of food. He died in his sleep in New York following a midnight steak meal.

Black, John Homer was born near Smith's Falls, Ontario in 1875. He taught school in Lanark and Frontenac Counties between 1892 and 1895. Black served as a C.P.R. telegraph operator and station agent until 1902 whereupon he became an auditor and assistant superintendent of the C.P.R.'s K.&P. Railway. His managerial abilities secured him the appointment as General Superintendent of the T.& N. O. Railway in 1904. Black, managed two northern Ontario power companies and became director of the Excelsior Insurance Company from 1916 to 1924 before being appointed vice-president of the Spruce Falls Paper Company in Kapuskasing.

Black, William Allan was born in Montreal in 1862. He worked on the Grand Trunk and Pacific Railways before moving to Manitoba in 1882. He joined the Ogilvie Milling Company the following year. In 1902 Black was appointed general manager of the company's western division. He was a member of he Winnipeg Board of Trade, councillor of the Winnipeg Grain and Produce Exchange and the director of an investment house, a coal mining firm, and a glass company. Black was the managing director of the Kaministiquia Power Company of Fort William.

Brooks, William F. was treasurer of the International Boom Company that was formed in 1904. Like Brooks, its officers were Minnesota men: Thomas

Shevlin was president, E.L. Carpenter was vice-president, G.S. Eddy was secretary, and Edward W. Backus, James A. Mathieu, and G.S. Parker were directors. Brooks entered into partnership with Edward Backus in 1899 forming the Backus Brooks Company. They started the Rainy River Lumber Company that operated until the partnership dissolved in 1910.

Buntin, Alexander (1822-1893) was born in Reton, Scotland and, at age 15, he sailed as an apprentice to Newfoundland. In Quebec City and Montreal, Buntin worked in the paper industry and learned the business of paper manufacturing and sales. In Hamilton, he worked for a paper goods wholesaler, William Miller and Company. Butin was involved in the company's construction of the Paper Manufactory in 1854 in Valleyfield, Quebec. It was a thoroughly modern mill with a 84-inch wide Fourdrinier machine and the mill employed 71 people. Alexander Buntin and Company was founded in 1854 and took over the paper mill. Two years later, Alexander's brother, James, joined the firm as a partner. Until about 1864, the Valleyfield factory was operated by the firm A. and J. Buntin. With the death of James in 1861, Alexander became the sole owner of the mill. By 1869, he introduced the first mechanical pulp process in North America. Around 1879, Buntin began to sell pulp to New England and the mill continued to be one of the world's most modern paper plants until about 1885.

Carpenter, Edmund A. (1863-1902) was the son and partner of W.H. Carpenter, one of the Lakehead's first sawmill operators. As a timber contractor he supplied ties and poles to the Port Arthur Electric Street Railway Co. and cut pulpwood which he exported using his own fleet of ships. Ill health forced him into an early retirement in the 1890s.

Carpenter, William Henry (1840-1900) arrived at the Lakehead in the early 1870s, just after the Dawson trail from the Lakehead to Red River had opened. Carpenter took over the route and operated it, transporting settlers and their goods to the west. The government took over the operation of the route in 1876 and Carpenter moved back to eastern Ontario. Returning to Fort William in 1881, he took over the holdings of Oliver-Davidson and owned two sawmills, one on the Kam River and the other on the Carp river. Poor health forced him to give up the business in 1888. He was subsequently appointed to the post of Sheriff of the District of Rainy River, a post he held til his death in 1900.

Carrick, John James (1873-1966) was born in Terre Haute, Indiana, grew up in Kincardine, Ont. and graduated from the University of Toronto in 1897. He came to Port Arthur in 1902-1903 as a real estate promoter and mine speculator. He also was part of the infamous "Timber Ring" headed by J.A. Little and including Don Hogarth and W.H. Russell. He served as mayor of Port Arthur and Conservative M.P.P. and M.P. for Port Arthur.

Cheesman, Al was a pioneer aviator at the Lakehead and was widely known in the mining fraternity. He contested an aldermanic seat in Port Arthur in the 1930s.

Clavet, George Onesime Philomene (1843-1909) moved to Prince Arthur's Landing from Silver Islet in 1875 to work as a grocery store clerk. He took an active part in the real estate boom in that town in the 1880s and expanded his grocery business in partnership with Marks and Dobie becoming one of the leading merchants at the Lakehead. At one time he owned the Northern Hotel, selling it in 1904 to Frank Mariaggi. Clavet served as mayor of Port Arthur from 1903-1907 and was many times an alderman.

Clergue, Francis (1856-1939) was an ebullient and hard-selling American entrepreneur. Born in Brewster, Maine, Clergue studied law at Maine State College and joined a local law firm in 1876. He was more interested in engineering and promotional activities than the law. Clergue was involved in the Bangor Street Railway development and the power station on the Penobscot River to supply it. Although an engineering success, the lack of capital resulted in the company's failure. From 1885 to 1895, Clergue was involved in many schemes: a resort hotel, a bank, a drydock, a railway and an electrical utility in Persia as well as mines in the American west. He was a talented promoter and an enthusiastic salesman, but his grand imaginative schemes were all poorly implemented. In 1902 at the height of its development, Clergue's integrated industrial complex at Sault Ste. Marie, Ontario was the largest self-contained industrial complex under the control of one man that Canada has ever seen. Clergue's Consolidated Lake Superior Corporation owned power, timber, transportation and subsidiary manufacturing. In 1903 all Sault operations ceased due to a lack of working capital. Clergue was ousted from the board of the company in 1908.

Cochrane, Frank (1852-1919) was born in Clarenceville, Lower Canada and became a Sudbury, Ontario merchant. In 1905 he was elected to the Ontario Legislature as a Conservative for the riding of East Nipissing. He became the Minister of Lands and Mines in the administration of James P. Whitney. Cochrane left Ontario provincial politics to become Minister of Railways and Canals in the federal government of Robert Borden.

Conmee, James (1848-1913) was born in Sydenham, Ontario where his father was engaged in the lumber trade. After serving for a time in the American Civil War, Conmee resumed his interest in lumbering and later became involved in mining and railroads. In the early 1870s he installed machinery in the Oliver-Davidson lumber mill. In 1880 he had a contract to cut timber along the C.P.R. line and he established a planing mill to manufacture lumber. His other businesses included the ownership of a hotel, setting up the first telephone communication in Port Arthur and constructing sections of the C.P.R., the P.A.D. & W., the C.N.R., and the Algoma Central

Railway. James Conmee was a promoter of the north's mining resources and was the first president of the Ontario Mining Institute in 1894. He was the second mayor of Port Arthur in 1885 and served six term as a Liberal in the Ontario Legislature and two terms in the House of Commons.

Cox, C.W. (1883-1958) was born near London, Ontario and arrived in Fort William in 1907 where he began work on the CPR. A flamboyant politician and lumberman, he dominated the Lakehead scene thoroughout the 1930s and 1940s. From 1934 to 1949, he served as both mayor of Port Arthur and M.P.P. for Port Arthur. He also represented Fort William as M.P.P. from 1948 to 1951.

The 4th Duke of Sutherland (b. 1851) was MP for Sutherland from 1872 to 1886 and served as the Mayor of Longton in 1895. He married Lady Millicent Fanny St. Clair-Erskine on October 20, 1884. She passed away in 1955.

Dawson, H.B. (1865-1941) was born in Peterborough and came to Port Arthur in 1886. Although he was a successful business man in both dry goods and sheet metal, he was mainly known for his pioneering work in public ownership. As chairman of the Public Utilities Commission (1934-1935), he was largely responsible for establishing publicly owned utilities in the city.

Egan, John (1848 - ?) was a superintendent for the West Division of the CPR.

Falls, Thomas was a tie and pulpwood contractor with an extensive operation in the Shabaqua area.

Ferguson, George Howard (1870-1946) was born Kemptville, Ontario and began to practice law in 1894. He was elected to the Ontario Legislature in 1905 as a Conservative and began a long career as a very popular politician. Ferguson took a special interest in the forest policy of Ontario and combined the office of Premier and Minister of Education. He resigned in 1930 to become the High Commissioner in London until 1935.

Finlayson, William was Ontario Minister of Lands and Forests in Howard Ferguson's government. Finlayson's 1927 Forestry Act classified provincial lands and was designed to promote reforestation. A five-member Forestry Board was established by Finlayson to assist in forest research, planning and management. Two years later the Provincial Forests Act extended the land classification process, created Provincial Forest Reserves and appointed a professional forester to promote good forestry practice and eventually develop sustainable forests. The election of Mitchell Hepburn government in 1934 and the Great Depression prevented the implementation of Finlayson's legislation.

Fortune, W.F. was a retailer of men's clothing, a grocer, and a land developer who, in 1911, was involved in the development of the subdivision near the CNR crossing on Fort William Road. He was also a well-known Liberal.

Gardner, Charles was a timber contractor for Provincial Paper.

Goodall, E. Lorne graduated from McGill University in 1924 as a mechanical engineer and joined the Abitibi Power and Paper Company. The following year he was transferred to Smooth Rock Falls as the engineer in charge of the construction of a high density bleaching plant. Goodall returned to McGill for three years of graduate studies in engineering related to paper mills. In 1930 he became the resident engineer at the Provincial Paper Company in Port Arthur during a period of mill modernization. Goodall left Abitibi in 1947 to become managing director and later vice-president of the Dryden Paper Company and he later served as director of the Dryden Paper Bag Company at Winnipeg.

Graham, George Alexander (1857-1927) came to Fort William in 1875 from St. Catharines, Ontario. In 1882, Graham formed a partnership with John T. Horne (1857-1927), a windjammer captain. The next year they built a planing mill on the west bank of the Kaministiquia River near the bascule bridge where the Ogilvie Flour Mill was later situated. In about 1892, a sawmill was added to the plant. Graham and Horne also acted as steamboat agents and owned the palace steamer, *Ocean*, a large schooner, *Sligo*, and the tug, *Salty Jack*. These two men proved to be shrewd and efficient businessmen who also operated a shipping and forwarding company that traded in coal, lime and salt. Graham also acted as a real estate developer of some magnitude in the early years of the twentieth century.

Greer, James (1874-1922), **Joseph** (1881-1941), **Charles H.** (1877-1938) were born in Caledon East and came to Murillo with their parents in 1882. The brothers were timber contractors providing ties to the Canadian Government Railway which was establishing a branch line to Lakehead elevators from the Grand Trunk Pacific Railway line. They built a saw mill in the Dog River Area to provide the ties, but when the project failed the mill was abandoned in 1913 after four years of operation. The Greers continued to supply ties and pilings for elevator construction. In 1924 the brothers established the Marshal Development Co. Ltd., a timber contracting business. C.H. Greer was called upon to testify before the Timber Comission in 1920.

Hazlewood, Richard Armstrong (1856-1928), partner of James Whalen.

Heenan, Peter (1873-1948) emigrated from Ireland to Kenora in 1902 where he worked as a locomotive engineer for the CPR. Through his links with the trade union movement, he entered politics as the Labour M.P.P. for Kenora. In 1926 he entered federal politics and served as Labour Minister under MacKenzie King. After returning to provincial politics, he was appointed Minister of Lands and Forests by Hepburn in 1934.

Hixon, F.P., from Lacrosse, Wisc., was a principal in the Gunflint and Lake Superior Railway Company. (See Arpin, Daniel J., above).

Hogarth, Donald M. (1878-1953) was born in Osecola, Ontario. He served in World War I as a Major-General and was mentioned in dispatches. Hogarth served in the Ontario Legislature from 1911-23 and from 1926-1934 and was a provincial Conservative party organizer. Primarily a mining magnate, he was involved with Steep Rock Iron Mine, Little Long Lake Iron Mine, and Marsden Red Lake Gold Mines to name a few. Hogarth was director of the Dominion Bank and had an interest in the Great Lakes Paper Company. Along with J. A. Little and J .J. Carrick, Hogarth was part of the "Old Tory Timber Ring."

Holt, Sir Herbert (1856-1941) was born in Dublin, Ireland where he studied civil engineering. He came to Canada in 1875 and worked on several Ontario and Quebec railways as an engineer. In 1883 and 1884, Holt was engineer and superintendent of construction on the prairie and mountain divisions of the C.P.R. He became a railway contractor completing several contracts in various parts of Canada between 1884 and 1892. In 1892, he entered the world of banking and finance and was the president of Sovereign Bank (1902-1904) and president of Royal Bank (1902-1934). By the early twentieth century, he was one of the leading financial figures in the nation. His interest in the Lakehead grew from his association with the Kam Power Company which generated hydro-electric power from Kakabeka Falls. He also had ties to many of Fort William's early iron and steel factories.

Howe, Clarence Decatur (1886-1960) was born in Waltham, Massachusetts and graduated from M.I.T. at age 21 in 1907. He taught civil engineering at Dalhousie University, Nova Scotia. In 1913 he took the position of Chief Engineer with the Board of Grain Commissioners where he supervised the construction of grain elevators. In 1916 Howe established C.D. Howe and Company of Port Arthur and pioneered innovative grain elevator building techniques at the Lakehead. In 1920 a group of New York investors hired Howe to construct a pulp mill in Port Arthur. The deal made Howe a director of the Kaministiquia Pulp and Paper Company. Howe was a few weeks late in completing the mill and the owners were late in making their payment due to a disastrous drop in the price of processed pulp from $100 to $25 a ton. Refinancing schemes were taking too long and Howe filed a mechanics lien against the Capital Trust Company. The resulting court case delivered ownership of the idle mill to the C.D. Howe Company. The mill was auctioned in 1922 to pay Howe for his construction costs. The new paper mill owner was Wisconsin industrialist, George Wilson Mead, president of the Consolidated Water Power and Paper Corporation.

Johnson, A.L. (1899-1990), brother of E.E. Johnson, was a graduate of West Point who had a distinguished career in the American Army during WWII. He attended the conferences at Yalta, Tehran and Quebec as part of the American delegation under Roosevelt.

Johnson, Edward Ellsworth (1890-1953) was born in Waupaca, Wisconsin and was educated in Wisconsin public schools and the University of Wisconsin where he received a law degree. After serving in World War I, as an American Army Captain, Johnson was made manager of the Pigeon River Lumber Company and came to Port Arthur in 1919 to wind up the company's affairs. From his living quarters above the company offices, he learned the logging business. He hired Oscar Lehtinen as his right hand man and learned to cruise timber, cook, drive a dog team and survive sub-zero weather. In 1925, Johnson established the Pigeon Timber Company and soon had a thousand loggers on the payroll. Logs were cut for domestic and foreign mills. Johnson supplied wood to the Great Lakes Paper Mill controlled by paper magnate E. Wellington Backus and Johnson became his assistant. The crash of 1929 wiped out the Backus empire and Johnson's personal fortune went with it. All he had left was his logging experience and his company which he then proceeded to rebuild. Using innovative approaches to build logging camps and modern machinery in his bush operation, Johnson was soon again a successful timber operator. In 1937 he acquired a fleet of vessels to carry pulpwood, iron ore and grain. He launched the Great Lakes Lumber and Shipping Mill in Fort William in 1940 to meet the wartime need for lumber.

Keefer, Francis Henry (Frank) (1860-1928) was born in Strathroy, Ont., and served as city solicitor and Conservative politician. He was a local booster credited with promoting a deep water seaway from the Atlantic to the Lakehead. Keefer terminal in Thunder Bay is named after him.

Kindleberger, Jacob (1875-1947) was born in Roumback, Alsace Lorraine, Germany and was educated at Ohio Wesleyan University. He was chairman of the board of the Kalamazoo Vegetable Parchment Company. Kindleberger served as the president of the Kalamazoo Chamber of Commerce three times and was a trustee of the Bronson Methodist Hospital and Ohio Wesleyan University.

King, John (bap. Jean Roy) 1858-1940, was born in Rimouski, Que. He arrived at the Town Plot (West Fort) in 1878 and worked as a CPR brakeman and on road construction. In 1905 he developed property at the corner of Victoria and Arthur Streets and, by 1911, had secured contracts for building stations and terminals along the railroad both east and west of the Lakehead. He was a supporter of the Liberal party and served as a Fort William alderman and as vice president of the Thunder Bay Historical Society, 1910-1911. King was appointed Vice Consul for Belgium in 1914.

King, Joseph Goodwin (1844-1910) emigrated, with his family, to Ontario in the 1860s. King worked in the flour milling industry in Port Hope and Keewatin before entering the grain business. He purchased the Port Arthur grain elevator from the C.P.R. in 1891 when Van Horne wanted to take the railway out of the grain trade. This wooden elevator built in 1883-84, was the first privately operated western terminal elevator. In 1892, to secure sufficient capital, King entered a five year agreement with the Marks family. Thomas Marks and his nephews took two thirds of the profits from the operation of this hospital elevator that treated wet and smutty grain for shipment. King was so successful in pioneering the first hospital elevator in Western Canada that he opened his own elevator in 1897 and adapted the best American grain-handling technology in his facility.

L'Abbe, Philip (1856-1941) was born in Luciville, Quebec and came to the Lakehead in 1880 where he worked as a timekeeper for the C.P.R. As an associate of mayor James Conmee, he began a successful business career with the purchase of the Brunswick Hotel (Cumberland & Lincoln) in 1884. L'Abbe, or Labby as he was also known, had a C.P.R. contract for cutting ties in Stanley. In 1897 he went to the Klondike Gold Rush, but returned in 1906 and embarked on business construction which included blocks on both Pearl and Court Streets.

Lehtinen, Oscar emigrated from Finland and arrived in the Lakehead via Duluth in 1911. After acquiring timber limits in Graham Township, he became a partner of E.E. "Eddie" Johnson in the Pigeon River Lumber Co. In 1925, they formed the Pigeon Timber Co. and, in 1930, the Great Lakes Shipping Co. with a sawmill, known as the Great Lakes Lumber Co. on the Mission River. The companies consolidated under the name Great Lakes Lumber and Shipping which operated until 1953. Lehtinen also owned and operated a large fur farm from 1930-1942.

Little, James Arthur (1868-1931) was Lt. Colonel and commanding officer of the 96th Lake Superior Regiment from 1909 to 1921. He was also a member of what came to be known as "the Old Tory Timber Ring." His confidantes were W. H. Russell, J. J. Carrick, former Port Arthur mayor and Conservative M.P. and M.P.P. and real estate promoter, as well as mining speculator and politician, Donald Hogarth. No timber was cut in Northwestern Ontario between 1911 and 1920 without involving the Timber Ring. It could provide pulp limits if a generous contribution to the Conservative party was made through Little or Hogarth. They were the district's political bosses and they controlled all government appointments to the Department of Land Forests and Mines and thereby controlled the business practices in these industries. Little, among other, was severely censured by the Timber Commission of 1920.

Lyons, James was born on a farm in Virginia, Ontario in 1878 where he remained until going to Sault Ste. Marie in 1900. He was superintendent of Algoma Steel from 1902 to 1904. Lyons started a fuel supply company and a builder's supply business at the Sault. After considerable experience as a city alderman and mayor of Sault Ste Marie, he was elected to the Ontario

Legislature in 1923.

Mackenzie, William (1849-1923) and **Donald Mann** (1853-1934) were railroad entrepreneurs who began assembling prairie railway lines in 1895. This formed the basis of the transcontinental Canadian Northern Railway which was completed in 1915. National and personal financial difficulties led to the nationalization of the railway in 1918. It later became part of the Canadian National Railway.

Mariaggi, Frank (?- 1918) was a native of Corsica who emigrated to Canada in 1872. He came to Port Arthur in 1904 and purchased the Northern Hotel (later renamed the Mariaggi Hotel on Water St.). In association with Carrick, he bought property at Bare Point in anticipation of railway expansion. When the venture failed, he returned, in 1906, to Corsica.

Marin, Captain Nicholas worked for the Silver Islet Mining Company as master of their steam yacht *Silver Spray*. It operated between 1872 and 1883 ferrying passengers and running errands. Marin was a safe and capable navigator with a great deal of experience and skill in working with tugs in Thunder Bay. As Captain of the *Laura Grace*, Marin towed logs for the Pigeon River Lumber Company for twenty years without an accident.

Marks, Thomas (1834-1900) was born in Ireland and emigrated to Upper Canada in the 1840s. He and his brother George opened a general store in 1868 at what would become Port Arthur. When the partnership ended, the Thomas Marks Company was founded in 1881. His business ventures were wide and varied, including the Northern Hotel, the steamship *Algonquin*, the Marks, King hospital elevator, the Port Arthur, Duluth and Western Railway as well as a sawmill. He also served as Mayor of Port Arthur in 1884 and was the town's major political influence (Conservative) during the period.

Mather, John (1827-1907) was born in Montrose, Scotland where he eventually managed a large sawmilling and shipbuilding firm. Gilmour and Company of Glasgow sent him to Lower Canada to manage their Ottawa and Gatineau rivers operations. Mather was an energetic and skilled woods and mill manager. He left the firm in 1876 to become the central figure in the development of the District of Keewatin. In 1879 Mather established the Keewatin Lumbering and Manufacturing Company at Rat Portage. By 1884, with the completion of the railway and the development of Winnipeg, the company was a financial success. Mather became a director of the Bank of Ottawa which was heavily involved in the lumber business. He was the vice-president of the Lake of the Woods Milling Company in 1887 and supervised the construction of this large flour mill in Keewatin. Mather also had interests in two Lake of the Woods area gold mines.

Mathieu, James A. (1869 -1966) was born in Alma, Wisconsin and came to Rainy River in 1902. In 1904, he took over the management of the Shevlin

Clark sawmill at Fort Frances until 1921. He formed J.A. Mathieu Ltd. in 1922 by purchasing and reorganizing the Border Lumber Co. Mathieu served as Conservative M.P.P. for Rainy River for 15 years between 1911 and 1929.

Matthews, George H. (1873-1953) was born in Moncton, New Brunswick and came to Port Arthur with his parents in 1884. After completing his high school education there, he worked at the Graham and Horne sawmill. He was employed briefly in a law office before becoming foreman for the Pigeon River Lumber Company and later was a construction foreman during the building of Elevator D in Fort William. Matthews used his experience in the lumber business to start the Matthews Sash and Door Company in 1903.

McCormick, Angus G., President of City Feed Company, Port Arthur.

McCuaig, John was a Port Arthur timber contractor specializing in piling timber and railway ties for the Canadian Northern Railway line from Port Arthur to Winnipeg. He was eventually hired by the Pigeon River Lumber Company to look after their labour requirements. He was a Glengarrian and had had considerable experience along the Ottawa River as a logger, but particularly as a raftsman. He was the typical "Man from Glengarry" portrayed so well by Ralph Connor in his famous novel of that title.

McDonald, Martin J. (1885-1953) was born in Appleton, Wisconsin and came to Port Arthur when he was 17. He worked for the Pigeon River Lumber Company and the Vigars-Shear Lumber Company. In 1910, McDonald established the Thunder Bay Lumber Company which absorbed Pigeon River Lumber and Vigars-Shears. This Port Arthur businessman later operated sawmills with George Wardrope (1899-1980). McDonald and Fred Brown launched the Steep Rock Lumber and Supply Company when the Steep Rock Iron Mine was being developed. George McDonald and J. Alexander "Sandy" McDonald were brothers of Martin J. McDonald all of whom began their business careers with Pigeon River Lumber. J.H. McLennan became a Port Arthur lumber dealer; Clarence "Sandy" Moors (1886-1965) became a feed and fuel dealer.

McMahon, Harry C. was a native of Douglas, Massachusetts and came to Canada as a young man. He lived in Gold Rock, Ontario before moving to Dryden where he lived for a time. He then moved to Fort William and was associated with the St. Louis Hotel. McMahon operated a hotel in the town of Mine Centre, Ontario until 1920 when he returned to the Lakehead. McMahon established the Port Arthur Beverage Company and was its president until he retired in 1948.

McPherson, George was a carriage rider in the Keewatin sawmill as well as a camp foreman during the winter and a log drive superintendent in the summer months. He was a veteran of World War I where he achieved the rank of Captain. Consequently, after the war, he became know as "Colonel"

McPherson or was addressed as the Kentucky Colonel. His mustache, his goatee and his fine clothes gave him the appearance of a southern gentleman.

Mead, George Houk (1877-1963) was born on Dayton, Ohio and received a B.Sc. from the Massachusetts Institute of Technology in 1900. Between 1897 and 1903, Mead worked for three different Ohio paper companies. In 1903 he was employed by the Artificial Silk Company of Philadelphia. The Mead Pulp and Paper Company of Dayton was organized in 1905 and George H. Mead was its president from 1915 to 1942. He also established G.H. Mead Company, the Mead Sales Company and the Mead Investment Company.

Mead, George Wilson (1871-1961) was a paper manufacturer and banker who was born in Chicago. Between 1894 and 1902 he developed water power and paper mills in Wisconsin Rapids, Wisconsin. In 1902, Mead was president of Consolidated Water Power and Paper Company. Two years later he completed the first entirely electrically motorized paper mill. Mead also served and mayor of Wisconsin Rapids.

Mellor, Charles (1872-1963) was born in Cheshire, England and first emigrated to Russia before coming to Port Arthur in 1904. In 1918, he established Mellor Timber Co. and later a feed business on S. Court. Street, Port Arthur. He also served on the Public Utilities Commission.

Morrison, Bruce (1881-196?) was a prospector who served one term (1934) as an alderman for Port Arthur.

Mooring, George (1865-1909) was a real estate developer who served as alderman (1891-92) and as president of the Board of Trade (1906) in Port Arthur. In 1908, he ran as an independent candidate in the provincial election. Mooring operated tie camps along the Arrow River.

Mosher, H.S., (1896-1979) was born in Elkhorn, Wisc., and lived in Port Arthur from 1927 to 1961. He served as Vice President and General Manager of the Newago Timber Company.

Murray, Jack ("Sand Bar") was a logging superintendent for Alger & Smith of Duluth. He had an interest in Pigeon River Lumber which he sold to cruise timber in Manitoba and Saskatchewan. In 1912 he built a sawmill in The Pas under the name Finger & Smith.

Nicholson, Harry (1859 -1949) arrived in Prince Arthur's Landing from Bruce Mines in 1874 at the age of 15. Among his many businesses was the operation of a ferry service to the Town Plot and a grocery store at the corner of Water and Lincoln Streets. In 1881 he started a dredging operation and from 1882-1896 he owned a clothing store on Water Street. During this period, he also worked as an employment agent providing men for the tie camps. Nicholson retired as a brakeman on the CPR in 1934 after 36 years of service.

Ogilvie, Shirley was born in Montreal in 1864 and was the son of the founder of the Ogilvie Milling Company. Ogilvie was sometimes the Ottawa agent for the Milling Company and, in 1902, lived in Montreal as secretary and director of the firm. In 1906 he was elected director of E.T. Bank and later held directorships of Consolidated Rubber and other businesses.

Oliver, Adam (1823-1882) and Joseph Davidson operated a sawmill at the Lakehead. It was established in the early 1870s and was situated on Island No. 1 on the south shore of the Kaministiquia River almost opposite the Hudson's Bay Company's Fort William. Adam Oliver, an Ingersoll, Ontario builder and contractor, went into partnership with Toronto lumberman, Joseph Davidson and Ingersoll business colleague, Peter Brown, in 1872. Their's was the first sawmill at Fort William. This moderate speed sawmill burned down in 1875. Oliver, Davidson and Brown were also prominent land speculators at the Lakehead in the 1870s and 1880s. Davidson was never a resident of Fort William. Adam Oliver continued in business until failing health caused him to retire.

Oliver, John A. (1871-1924) Mayor of Port Arthur 1913, 1914; alderman 1910-11; president of the Board of Trade 1916-17. Oliver came to the Lakehead in the 1890s to work as a lumber camp manager for Ed Carpenter. His real talent rested, however, in politics. His efforts on behalf of the Conservative party earned him the appointment of Crown Timber Agent for the Port Arthur district in 1905. Strongly attacked by the Timber Commission in 1920, he retired his post in 1921. His term as mayor coincided with Port Arthur's major real estate boom.

Paterson, Norman McLeod (1883-1983) was born in Portage la Prairie, Manitoba. He worked as a telegraph operator, private secretary, chief clerk and purchasing agent for the Canadian Northern Railway Company. In 1903 he entered the grain business with his father, H.S. Paterson. His own firm, N.M. Paterson Company Limited, was established at Fort William in 1908. It eventually controlled and operated 109 elevators between Fort William and Saskatchewan and a four million bushel terminal elevator at Fort William. Paterson's business interests included: thirty-one lake freighters, engineering services, a lumber company, electrical power, a marine railway and drydock company, and a grain insurance company. He was appointed to the Senate in 1940. Paterson was also noted for his philanthropy in the arts, education and recreation.

Peters, R.J. was a prominent financier, railwayman, industrialist and lumber manufacturer from Manistee, Michigan. The focus of his business activity was Michigan, but he had vast mine and railway interests across the continent. In 1884, he bribed some Ontario politicians in order to secure favourable timber legislation. He had purchased 10,000 acres of land on the Canadian side of the Pigeon River and was angered when the Oliver Mowat government would not give him a patent to cut and export the pine stands there. Peters also became involved in mining in Ontario when he purchased the Beaver Mine developed by the Oliver Daunais Group (T.A. Keefer, Daniel McPhee, W.H. Furlong). Peters owned one of the finest yachts on Lake Michigan which was named the "Sigma" after one of his railway stations. In 1912, purely by coincidence, this fine vessel was bought by James Whalen of Port Arthur and adorned the lake front of the two cities. It took many local citizens on cruise trips on the bay until the First World War when it was dismantled.

Russell, Walter H. (1882-1954) was born in Plymouth New Hampshire and was a graduate of Dartmouth College in New Hampshire and the University of Michigan law school in 1906. The young Detroit lawyer came to Lakehead and started the Russell Timber Company and became a pulp wood exporter. In 1913, he promoted a plan to build a pulp mill on the Current River in Port Arthur with financial support of the municipality. He was backed by local Conservative politicians, Mayor S.W. Ray, M.P. J.J. Carrick and M.P.P. Donald Hogarth. Russell also had strong ties to Ontario Conservatives. The pulp mill was opposed by the Port Arthur Parks Board composed mostly of Liberals. Russell's scheme to bonus the plant with public money went down to defeat in part because the public objected to turning Current River into an industrial site. Russell was a member of the Conservative Association and was appointed police court magistrate for Port Arthur and district in 1945.

Schram, Oliver owned a sawmill at Hymers along the P.A.D. & W. line.

Scott, William first came to the Thunder Bay area with D.J. Arpin on the steamer *Dixon* in 1898. They landed at Grand Portage to look at timber lands. The following year, Scott and Arpin purchased timber and interested F.P. Hixon and H. Finger in organizing the Pigeon River Lumber Company. They operated the mill, purchased from Graham and Horne, for one year and then moved to Port Arthur to a new saw and planing mill which opened in 1902.

Shear, Captain Herbert was a local entrepreneur and in 1889 was the superintendent of the Badger Mine near Silver Mountain.

Sheppard, W.J. (aka Shepherd) was an Orillia, Ontario lumberman who saw the potential of the falls on the Spanish River for the development of hydro electric power and the establishment of a paper mill. The firm, Spanish River Pulp and Power, was formed with Sheppard as president. M.J. Dodge of New York was the vice-president and was said to be the "the power behind the throne." It was Dodge who was granted timber limits in 1900 by the Ontario government. Sheppard and Dodge had worked together before; Dodge was the vice-president of Sheppard's Georgian Bay Lumber Company.

Smith, Erle (1890-1976) was born in West Fort William and attended Port Arthur High School before pursuing a career in civil engineering. In 1907 he worked as a civil engineer for the City of Port Arthur. Smith joined C.D. Howe and Company in 1916 and supervised the construction of grain elevators and pulp mills. In 1931, he began working for the Department of Northern Development (highways) and three years later was made District Engineer.

Stewart, Alexander (1829-1912) was born in York County, New Brunswick. In 1849, he moved to Wisconsin and settled where the city of Wausau is now located and worked in the lumber business. He became involved in Republican politics in 1884 and was elected to Congress twice and served there from 1895 to 1901.

Styffe, Oscar (1885-1943) was born in Lossekopp, Norway where he was educated in forest management. He emigrated to Michigan and moved on to Port Arthur in 1911. As a moderately successful timber contractor, he came into conflict with Charlie Cox in both the business and political arena. Styffe served as a Port Arthur alderman in 1935 and was appointed Norwegian consul in 1941.

Sweezy, Robert Oliver was a consulting engineer specializing in water power, pulp and paper, timber lands and financial reports. Educated in private schools in Chicotimi and Queen's University, he was a member of the Canadian Institute of Metallurgy, and president of Newman, Sweezy Company, Montreal investment bankers. Sweezy explored resource and timber lands of Northern Ontario and Quebec and other parts of Canada and was a lumberman and timber operator. In addition, Sweezy was involved in the location and construction work for the Canadian Northern Railway as a consulting engineer and also did consulting work for various investment houses and public corporations. He entered business on his own in 1921.

Taschereau, Louis Alexandre (1867-1952) was born at Quebec and educated there at Laval. He was called to the bar in 1889. Taschereau was elected to represent Montmorency as a Liberal in the Legislative Assembly of Quebec in 1900. In 1907 he was appointed Minister of Public Works and later became Premier of Quebec. He served his province until 1936.

Tracey, Matt was an Ottawa Valley veteran timber man and pioneer citizen of the Lakehead. He was entrusted with the operations at the mouth of the Pigeon River.

Tudhope, James Brockett was born of Scottish ancestry in the Township of Oro in Simcoe County, Ontario in 1858. Like his friend, W. J. Sheppard, Tudhope was an Orillia businessman. He was vice-president of the Universal Signal Company and became the head of Tudhope Carriage Factory Co. in Orillia along with the Tudhope Motor Co. and Carriage Factories. Tudhope served as secretary-treasurer of the Spanish River Pulp and Paper Company. He held several political positions becoming councillor, reeve and mayor of

Orillia before being elected as the Liberal representative of East Simcoe in 1910.

Van Horne, Sir William Cornelius (1843-1915) was born in Chelsea, Illinois and his railway career began on American railways at age 14. By 1880, he was a general superintendent and in 1882 became general manager of the Canadian Pacific Railway. He was a man of great determination and stamina as well as being a brilliant manager and a man of refined taste and intellectual curiosity.

Vigars, Richard (1848-1930) was born in Devonshire, England and settled in Bowmanville in 1859. He brought his family and his brother, William, to Prince Arthur's Landing in 1875, after a brief stay in Marquette, Michigan. Richard was in the building and general contracting profession and, in 1884, entered the lumbering business with his brother. They organized the Vigars Brothers Company in 1886 and built a lumber and planing mill on or near the site of the Comnee sawmill. They eventually joined forces with the Northern Land and Lumber Company. The Vigars sawmill was also the site of the Port Arthur Light and Power Company's steam-driven electric lighting plant and was destroyed by fire in August 1894. Vigars became Port Arthur's mayor in 1905 and served as a councillor, a member of the Board of Health, a public school trustee and as president of the Board of Trade (1913).

Wardrope, George C. (1899-1980) was born in Montreal and raised in Ottawa, Belleville and London. He established an insurance and real estate firm following the First World War, managed a contracting company in the 1920s and a lumber supply company in the early 1940s. As a Conservative, Wardrope ran unsuccessfully against C.D. Howe in the election of 1935, but gained the Port Arthur seat for Port Arthur provincially in 1951, holding it until 1967. He served as Minister of Reform Institutions (1958-1961) and as Minister of Mines (1961-1967). Prior to that, he served as alderman in Port Arthur for ten years and then briefly as an alderman in Thunder Bay, resigning in 1970.

Wark, Doug was a native of New Brunswick who had gone to Wisconsin and became associated with Herman Finger. For fifteen years, Wark was Finger's timber cruiser and superintendent of woods operations. Wark and Finger laid out two large logging camps located one mile apart near the mouth of the Pigeon River. Doug Wark was in control of the entire woods operation and supervised the foreman who was in charge of Camps 1 and 2.

Whalen, James (1869-1929) was born in Collingwood and came to Prince Arthur's Landing in 1873 with his parents. He was educated in local public schools and entered the logging business and was busy taking out spruce pulpwood from the Pic River in 1896. Whalen married Laurel T. Conmee, the daughter of James Conmee. Soon he was involved in railway contracting and a wide variety of business enterprises due in part to his association with his politician father-in-law. He built the Lyceum Theatre and the Whalen Building. He helped to create what became the Port Arthur Shipbuilding Company, and established the Canadian Towing and Wrecking Company, General Realty Corporation, Whalen Pulp and Paper Limited and the Canadian West Coast Navigation Company.

Whitney, Sir James P. (1843-1914) was first elected to provincial office in 1888 and was chosen leader of the Conservative party in 1896. He rebuilt the party into a formidable political machine which won the general election of 1905. He was Premier of Ontario from then until his death in 1914. His administration was noted for the establishment of publically-owned hydroelectric power, workmen's compensation legislation, reforms to the Ontario Railway and Municipal Board, changes to the use of French in public school, and control of the liquor trade.

Appendix 2

Index of Pulp and Paper Mills
in Northwestern Ontario

The Dryden Mill was purchased by Anglo-Canadian in 1964 and subsequently sold to Reed Paper International of London, England in 1968. Great Lakes Paper (now Avenor) bought the mill in 1979 and upgraded the facility which presently employs 920 persons producing 310,000 metric tonnes of white paper and 80,000 tonnes of pulp per year.

The Espanola Mill was sold by KVP to E.B. Eddy Forest Products Ltd.

The Fort Frances and Kenora Mills were sold by O&M to Boise Cascade Corporation in 1968. In 1993, the mills became publicly owned under the name Rainy River Forest Products. Stone-Consolidated Corporation purchased the mills in 1995. Stone-Consolidated merged with Abitibi-Price to form Abitibi Consolidated in 1997.

The Fort William (Mission) Mill is presently up for sale by Abitibi Consolidated. The mill produces 150,000 tonnes/year of 12% recycled pulp and 310,000 tonnes/year of newsprint and newsprint specialties.

The Great Lakes Paper Company went into receivership in 1931, but continued to operate partially. In 1936, the company was completely reorganized under government charter as The Great Lakes Paper Company Ltd. Canadian Pacific Investments acquired 50.4% of the company in 1979. After the merger between CIP (Canadian International Paper Co.) and GLP in 1988, the company was renamed Canadian Pacific Forest Products Ltd. In 1993, Canadian Pacific Enterprises divested its shares of CPFP creating an independent forest products company called Avenor. Avenor's Thunder Bay facility currently employs 850 in the newsprint mill producing 507,000 metric tonnes/year of newsprint and 653 in the pulp mill producing 460,000 metric tonnes/of pulp.

The Iroquois Falls Mill remains an Abitibi Consolidated mill and presently produces 310,000 tonnes/year of newprint and newsprint specialties.

Kimberly-Clark, Terrace Bay, emerged from the Pulpwood Supply Co., a consortium of American paper companies which acquired pulpwood limits north of Lake Superior. In 1947 the mill, Long Lake Pulp and Paper Co., and the townsite of Terrace Bay, were built. In 1958 the name of the mill was changed to Kimberly-Clark Pulp and Paper Co.

The Lake Sulphite plant was built in 1937 at Red Rock, but went into receivership in 1938. In 1941 Brompton Pulp and Paper of Quebec took over the property, but did not start up the mill until 1945. St. Lawrence Corporation took over the mill in the 1950's and, in 1957, sold it to Domtar. The mill was completely converted from newsprint production to linerboard in 1992.

The Marathon Paper Mill was built in 1945 at Marathon (Peninsula Harbour) and was bought by American Can in 1957. In 1983, 80% of the operation was sold to James River Corporation, a Virginia based company, and 20% to Buchanan Forests Products, Thunder Bay. The company is presently merging with the Fort Howard Corporation under the new title Fort James Corporation. At present, the Marathon operation has a production capacity of 353 tonnes/day of bleached kraft pulp.

The Provincial Mill (Port Arthur) was sold by Abitibi Price to the employees on June 1, 1993. In 1997, the company was purchased by Roland Inc. The present production capacity of the mill is 160,000 tonnes/year of machine coated paper.

The Sault Ste. Marie Mill was sold by Abitibi to St. Mary's Paper (Chicago) in 1984. The mill went into receivership in 1993, but continues to operate while looking for a buyer. The mill produces 600 tonnes/day of supercalendered groundwood and printing and specialty papers.

The Spruce Falls Mill (Kapuskasing) was sold to the employees by Kimberly-Clark in 1991. The facility was 48% employee owned, 48% owned by Tembec and 4% owned by the community until 1997 when Tembec bought out the employees.

The Smooth Rock Falls Mill was purchased from Abitibi by Mallette Kraft Pulp and Paper and sold to Tembec Inc. in 1995. The mill produces northern softwood bleached kraft pulp.

The Sturgeon Falls Mill is owned by MacMillan Bloedel Ltd. and employs 170 persons. It produces 25 tons/day of 100% recycled corrugated medium used in cardboard.

The Thunder Bay Mill was sold by Consolidated Water Power and Paper Co. to the Thunder Bay Paper Co. in 1921 and then to Abitibi in 1928. The mill was closed by Abitibi in 1993, but later reopened as Thunder Bay Packaging Inc., a subsidiary of St. Laurent Paper Board Inc.

Index